Lial Video Workbook

Christine Verity

Prealgebra

Sixth Edition

Margaret L. Lial
American River College

Diana L. Hestwood
Minneapolis Community and Technical College

The author and publisher of this book have used their best efforts in preparing this book. These efforts include the development, research, and testing of the theories and programs to determine their effectiveness. The author and publisher make no warranty of any kind, expressed or implied, with regard to these programs or the documentation contained in this book. The author and publisher shall not be liable in any event for incidental or consequential damages in connection with, or arising out of, the furnishing, performance, or use of these programs.

Reproduced by Pearson from electronic files supplied by the author.

Copyright © 2018 Pearson Education, Inc.
Publishing as Pearson, 330 Hudson Street, NY NY 10013

All rights reserved. No part of this publication may be reproduced, stored in a retrieval system, or transmitted, in any form or by any means, electronic, mechanical, photocopying, recording, or otherwise, without the prior written permission of the publisher. Printed in the United States of America.

2 2020

ISBN-13: 978-0-13-454098-6
ISBN-10: 0-13-454098-0

CONTENTS

CHAPTER 1	INTRODUCTION TO ALGEBRA: INTEGERS	1
CHAPTER 2	UNDERSTANDING VARIABLES AND SOLVING EQUATIONS	33
CHAPTER 3	SOLVING APPICATION PROBLEMS	59
CHAPTER 4	RATIONAL NUMBERS: POSITIVE AND NEGATIVE FRACTIONS	81
CHAPTER 5	RATIONAL NUMBERS: POSITIVE AND NEGATIVE DECIMALS	131
CHAPTER 6	RATIO, PROPORTION, AND LINE/ANGLE/TRIANGLE RELATIONSHIPS	183
CHAPTER 7	PERCENT	219
CHAPTER 8	MEASUREMENT	247
CHAPTER 9	GRAPHS AND GRAPHING	269
CHAPTER 10	EXPONENTS AND POLYNOMIALS	297
ANSWERS		321

Name: Date:
Instructor: Section:

Chapter 1 INTRODUCTION TO ALGEBRA: INTEGERS

1.1 Place Value

Learning Objectives
1   Identify whole numbers.
2   Identify the place value of a digit through hundred-trillions.
3   Write a whole number in words or digits.

Key Terms

Use the vocabulary terms listed below to complete each statement in exercises 1−3.

       place value system   digits   whole numbers

1. The ten _____ in our number system are 0, 1,2 3, 4, 5, 6, 7, 8, and 9.

2. The _____ are 0, 1, 2, 3, 4, and so on.

3. A _____ is a number system in which the location, or place, where a digit is written gives it a different value.

Objective 1   Identify whole numbers.

Video Examples

Review this example for Objective 1:
1. Identify the whole numbers in this list.

   $67, -3, 200, 1.9, \frac{8}{9}, 0, 0.333, 8\frac{3}{4}, 4$

   The whole numbers are 0, 4, 67, and 200.

Now Try:
1. Identify the whole numbers in this list.

   $59, -6, 350, 9.7, \frac{9}{8}, 0, 7\frac{1}{3}, 0.999, 5$

   _____

Objective 1 Practice Exercises

For extra help, see Example 1 on page 5 of your text.

Choose the whole numbers in each set of numbers.

1. $-\frac{3}{4}, 2, 1.398, -2\frac{1}{2}, -1.04, -2, \frac{1}{5}, -2.6428$ 1. _____

2. $-0.5, -1, 4, 3, -4.87, -22, \frac{4}{3}, 1\frac{1}{2}$ 2. _____

3. $2.718, -1, 365, 22.4, -6.02, 1, -4\frac{2}{5}$ 3. _____

Name: Date:
Instructor: Section:

Objective 2   Identify the place value of a digit through hundred-trillions.

Video Examples

Review this example for Objective 2:
2. Identify the place of each 5 in the number.
    7,946,145,057

    7,946,145,057
            ↑ ↑ tens place
    thousands place

Now Try:
2. Identify the place of each 3 in the number.
    9,854,310,327

    _____

Objective 2 Practice Exercises

For extra help, see Example 2 on page 5 of your text.

Give the place value of the digit 6 in each number.

4. 639,111,192

5. 94,164,372,757

6. 6458

4. _____

5. _____

6. _____

Objective 3   Write a whole number in words or digits

Video Examples

Review these examples for Objective 3:
3.
   a. Write 7,096,381 in words.

   seven million, ninety-six thousand, three hundred eighty-one

   b. Write 41,677,000,500,000 in words.

   forty-one trillion, six hundred seventy-seven billion, five hundred thousand

4. Write each number using digits.

   a. Six hundred forty-two thousand, ten

   The answer is 642,010.

   b. Eighty-nine billion, twenty-five thousand, six hundred

   The answer is 89,000,025,600.

Now Try:
3.
   a. Write 9,075,862 in words.

   _____

   b. Write 54,800,543,700,100 in words.

   _____

4. Write each number using digits.

   a. Seven hundred fifty-three thousand, six

   _____

   b. Eleven billion, ten thousand, twenty-five

   _____

Name: Date:
Instructor: Section:

Objective 3 Practice Exercises

For extra help, see Examples 3–4 on pages 5–6 of your text.

Write each number in words.

7. 59,504,806,873 7. _____

8. 9,671,000,637 8. _____

Write the number using digits.

9. Nine hundred eighty-seven million, three hundred thirty 9. _____

Name: Date:
Instructor: Section:

Chapter 1 INTRODUCTION TO ALGEBRA: INTEGERS

1.2 Introduction to Integers

Learning Objectives
1 Write positive and negative numbers used in everyday situations.
2 Graph numbers on a number line.
3 Use the < and > symbols to compare integers.
4 Find the absolute value of integers.

Key Terms

Use the vocabulary terms listed below to complete each statement in exercises 1–3.

    number line   **integers**   **absolute value**

1. The _____ of a number is its distance from 0 on the number line.

2. A _____ is used to show how numbers relate to each other.

3. The whole numbers together with their opposites and 0 are called _____.

Objective 1   Write positive and negative numbers used in everyday situations.

Video Examples

Review these examples for Objective 1:	Now Try:
1. Write each negative number with a negative sign. Write each positive number in two ways. **a.** The Dead Sea is 1299 feet below sea level. The answer is –1299 ft. **b.** The mountain had elevation of 17,000 feet. The answer is +17,000 feet or 17,000 feet.	1. Write each negative number with a negative sign. Write each positive number in two ways. **a.** A diver descends to a depth of 150 ft below sea level. _____ **b.** The temperature rose to 68°F. _____

Name: Date:
Instructor: Section:

Objective 1 Practice Exercises

For extra help, see Example 1 on page 14 of your text.

Write a signed number for each of the following.

1. Following a defeat, an army retreats 46 kilometers.   1. _____

2. The company had a profit of $830.   2. _____

3. A corporation has a shortfall of $1.2 million   3. _____

Objective 2   Graph numbers on a number line.

Video Examples

Review this example for Objective 2:

2. Graph each number on the number line.

$$-\frac{1}{2}, -3, -\frac{5}{2}, \frac{1}{4}, 1\frac{7}{8}, 3$$

Draw a dot at the correct location for each number.

$$-3, -\frac{5}{2}, -\frac{1}{2}, \frac{1}{4}, 1\frac{7}{8}, 3$$

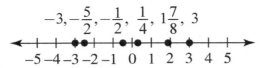

Now Try:

2. Graph each number on the number line.

$$-3, -5, -1\frac{1}{2}, \frac{2}{3}, 3.5, 4$$

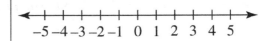

Objective 2 Practice Exercises

For extra help, see Example 2 on page 14 of your text.

Graph each set of numbers on a number line.

4. $-2, -1, 0, 1, 2, 5$

4.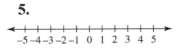

5. $-4.5, -1.5, -0.5, 0, 1.5, 2.5$

5.

6. $4\frac{1}{2}, 1, -2\frac{3}{4}, -1\frac{1}{2}, -\frac{1}{4}$

6.

Name: Date:
Instructor: Section:

Objective 3 Use the < and > symbols to compare integers.

Video Examples

Review these examples for Objective 3:

3. Write < or > between each pair of integers to make a true statement.

   a. 0 _____ 7

   0 is to the left of 7 on the number line, so 0 is less than 7. Write 0 < 7.

   b. −2 _____ −5

   −2 is to the right of −5, so −2 is greater than −5. Write −2 > −5.

   c. −6 _____ 7

   −6 is to the left of 7, so −6 is less than 7. Write −6 < 7.

Now Try:

3. Write < or > between each pair of integers to make a true statement.

   a. 0 _____ −8

   _____

   b. −3 _____ −9

   _____

   c. 11 _____ −4

   _____

Objective 3 Practice Exercises

For extra help, see Example 3 on page 15 of your text.

Write < or > in each blank to make a true statement.

7. −23 _____ −32

8. −6 _____ 0

9. −5 _____ −3

7. _____

8. _____

9. _____

Objective 4 Find the absolute value of integers.

Video Examples

Review these examples for Objective 4:

4. Find each absolute value.

   a. $|3|$

   The distance from 0 to 3 on the number line is 3 spaces. So, $|3| = 3$.

   b. $|-3|$

   The distance from 0 to −3 on the number line is 3 spaces. So, $|-3| = 3$.

Now Try:

4. Find each absolute value.

   a. $|12|$

   _____

   b. $|-13|$

   _____

Name: Date:
Instructor: Section:

c. $|0|$

$|0| = 0$ because the distance from 0 to 0 on the number line is 0 spaces.

c. $|0|$

_____

Objective 4 Practice Exercises

For extra help, see Example 4 on page 16 of your text.

Find each absolute value.

10. $|-95|$

10. _____

11. $|21|$

11. _____

12. $|-788|$

12. _____

Name: Date:
Instructor: Section:

Chapter 1 INTRODUCTION TO ALGEBRA: INTEGERS

1.3   Adding Integers

Learning Objectives
1   Add integers.
2   Identify properties of addition.

Key Terms

Use the vocabulary terms listed below to complete each statement in exercises 1−5.

addends   sum   addition property of 0

commutative property of addition

associative property of addition

1. By the_____, changing the order of the addends in an addition problem does not change the sum.

2. In addition, the numbers being added are called the _____.

3. By the_____, changing the grouping of the addends in an addition problem does not change the sum.

4. The answer to an addition problem is called the _____.

5. The _____ says that adding 0 to any number leaves the number unchanged.

Objective 1   Add integers.

Video Examples

Review these examples for Objective 1:
1. Use a number line to find $(-3)+(-2)$.

    Start at 0. Move 3 places to the left. Then move 2 places to the left.

    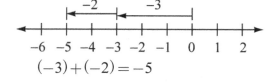

    $(-3)+(-2)=-5$

Now Try:
1. Use a number line to find $(-4)+(-1)$.

   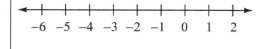

   _____

Name: Date:
Instructor: Section:

2. Add.

   a. $-9+(-4)$

   Step 1 Add the absolute values.
      $|-9|=9$ and $|-4|=4$
   Add $9+4$ to get 13.
   Step 2 Use the common sign as the sign of the sum. Both numbers are negative, so the sum is negative.
      $-9+(-4)=-13$

   b. $7+8$

   Both numbers are positive, therefore the sum is positive.
      $7+8=15$

3. Add.

   a. $-7+4$

   Step 1 $|-7|=7$ and $|4|=4$
      Subtract $7-4$ to get 3.
   Step 2 -7 has the greater absolute value and is negative, so the sum is also negative.
      $-7+4=-3$

   b. $-6+15$

   Step 1 $|-6|=6$ and $|15|=15$
      Subtract $15-6$ to get 9.
   Step 2 15 has the greater absolute value and is positive, so the sum is also positive.
      $-6+15=9$

2. Add.

   a. $(-25)+(-11)$

      _____

   b. $19+21$

      _____

3. Add.

   a. $-19+3$

      _____

   b. $-4+25$

      _____

Objective 1 Practice Exercises

For extra help, see Examples 1–4 on pages 21–23 of your text.

Add by using a number line.

   1. $-8+5$ **1.** _____

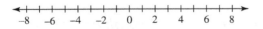

Add.

   2. $7+(-22)$ **2.** _____

Name: Date:
Instructor: Section:

Write an addition problem for the situation and find the sum.

3. While playing a card game, Diane first gained 23 points, then lost 16 points, and finally gained 11 points. What was her final score?

3. _____

Objective 2   Identify properties of addition.

Video Examples

Review these examples for Objective 2:

6. In each addition problem, pick out the two addends that would be easiest to add. Write parentheses around those addends. Then find the sum.

   a. $-10+10+7$

   Group $-10+10$ because the sum is 0.
   $(-10+10)+7$
   $\quad\;\; 0 \;\;\; +7$
   $\quad\quad\;\; 7$

   b. $-35+16+4$

   Group $16+4$ because the sum is 20, which is a multiple of 10.
   $-35+(16+4)$
   $-35+\;\;\; 20$
   $\;\;-15$

Now Try:

6. In each addition problem, pick out the two addends that would be easiest to add. Write parentheses around those addends. Then find the sum.

   a. $-20+19+(-19)$

   _____

   b. $8+2+(-17)$

   _____

Objective 2 Practice Exercises

For extra help, see Examples 5–6 on pages 24–25 of your text.

Rewrite each sum using the commutative property of addition, and find the sum each way.

4. $-6+13$

4. _____

In each addition problem, write parentheses around the two addends that would be easiest to add. Then find the sum.

5. $8+(-15)+(-5)$

5. _____

6. $-11+11+2$

6. _____

Name: Date:
Instructor: Section:

Chapter 1 INTRODUCTION TO ALGEBRA: INTEGERS

1.4   Subtracting Integers

Learning Objectives
1   Find the opposite of an integer.
2   Subtract integers.
3   Combine adding and subtracting of integers.

Key Terms

Use the vocabulary terms listed below to complete each statement in exercises 1−2.

opposite   additive inverse

1. The _____ of a number is the same distance from 0 on the number line as the original number, but located on the other side of 0.

2. The opposite of a number is its _____.

Objective 1   Find the opposite of an integer.

Video Examples

Review this example for Objective 1:	Now Try:

1a. Find the opposite (additive inverse) of the number. Show that the sum of the number and its opposite is 0.
 	
  4
 	
The opposite of 4 is −4 and $4+(-4)=0$.
	1a. Find the opposite (additive inverse) of the number. Show that the sum of the number and its opposite is 0.
	
	  25
	
	  _____

Objective 1 Practice Exercises

For extra help, see Example 1 on page 30 of your text.

Find the opposite (additive inverse) of each number. Show that the sum of the number and its opposite is 0.

1.   −11

2.   4

1. _____
    _____

2. _____
    _____

Name: Date:
Instructor: Section:

3. −5

3. _____

   _____

Objective 2 Subtract integers.

Video Examples

Review these examples for Objective 2:

2. Make two pencil strokes to change each subtraction problem into an addition problem. Then find the sum.

   a. $6 - 9$

   Change 9 to −9. Change subtraction to addition.
   $6 - 9 = 6 + (-9) = -3$

   d. $-3 - 12$

   Change 12 to −12. Change subtraction to addition.
   $-3 - 12 = -3 + (-12) = -15$

   b. $-8 - (-5)$

   Change −5 to 5. Change subtraction to addition.
   $-8 - (-5) = -8 + 5 = -3$

   c. $4 - (-9)$

   Change −9 to 9. Change subtraction to addition.
   $4 - (-9) = 4 + 9 = 13$

Now Try:

2. Make two pencil strokes to change each subtraction problem into an addition problem. Then find the sum.

   a. $13 - 15$

   _____

   d. $-8 - 16$

   _____

   b. $-10 - (-9)$

   _____

   c. $12 - (-11)$

   _____

Objective 2 Practice Exercises

For extra help, see Example 2 on page 31 of your text.

Subtract by changing subtraction to addition.

4. $-2 - (-11)$

4. _____

5. $-13 - 1$

5. _____

6. $12 - (-5)$

6. _____

Name: Date:
Instructor: Section:

Objective 3 Combine adding and subtracting of integers.

Video Examples

Review this example for Objective 3:

3. Simplify by completing all the calculations.
   $-6-15-18+3$

   Change all subtractions to adding the opposite. Change 15 to –15. Change 18 to –18. Then add from left to right.

   $-6 - 15 - 18 + 3$
   $-6+(-15)+(-18)+3$
   $-21 + (-18)+3$
   $-39 +3$
   -36

Now Try:

3. Simplify by completing all the calculations.
   $-9-14-21+6$

   _____

Objective 3 Practice Exercises

For extra help, see Example 3 on page 31 of your text.

Simplify.

7. $-2-16-(-18)$ 7. _____

8. $13-(-8)+(-7)-7$ 8. _____

9. $-14-(-3)-0+(-11)$ 9. _____

Name: Date:
Instructor: Section:

Chapter 1 INTRODUCTION TO ALGEBRA: INTEGERS

1.5 Problem Solving: Rounding and Estimating

Learning Objectives
1. Locate the place to which a number is to be rounded.
2. Round integers.
3. Use front end rounding to estimate answers in addition and subtraction.

Key Terms

Use the vocabulary terms listed below to complete each statement in exercises 1−3.

    rounding   estimate   front end rounding

1. _____ is rounding to the highest possible place so that all the digits become zeros except the first one.

2. In order to find a number that is close to the original number, but easier to work with, use a process called _____.

3. _____ to find an answer close to the exact answer.

Objective 1 Locate the place to which a number is to be rounded.

Video Examples

Review this example for Objective 1:	**Now Try:**
1b. Locate and draw a line under the place to which the number is to be rounded. Then answer the question. Round $573 to the nearest hundred. Is $573 closer to $500 or $600? 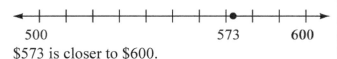 500 573 600 $5̲73 is closer to $600.	**1b.** Locate and draw a line under the place to which the number is to be rounded. Then answer the question. Round $642 to the nearest hundred. Is $642 closer to $600 or $700? _____

Objective 1 Practice Exercises

For extra help, see Example 1 on page 34 of your text.

Locate the place to which the number is rounded by writing the appropriate digit.

1. −135   Nearest ten   1. _____

2. −99,102   Nearest hundred   2. _____

3. 645,371   Nearest ten-thousand   3. _____

Name: Date:
Instructor: Section:

Objective 2   Round integers.

Video Examples

Review these examples for Objective 2:

4a. Round −5623 to the nearest ten.

Step 1 −56<u>2</u>3
           ten's place

Step 2 The next digit to the right is 3, which is 4 or less.
           −56<u>2</u>3
              Leave 2 as 2.

Step 3 −56<u>2</u>3 rounds to −5620.

5a. Round −86,732 to the nearest ten-thousand.

Step 1 −<u>8</u>6,732

Step 2 The next digit to the right is 6, which is 5 or more.
           −<u>8</u>6,732
              Change 8 to 9.

Step 3 −<u>8</u>6,732 rounds to −90,000.

Now Try:

4a. Round −4738 to the nearest ten.

_____

5a. Round −75,493 to the nearest ten-thousand.

_____

Objective 2 Practice Exercises

For extra help, see Examples 2–5 on pages 35–37 of your text.

Round each number to the indicated place.

4.   −6694 to the nearest thousand   4. _____

5.   814,118 to the nearest ten thousand   5. _____

6.   −4,697,134 to the nearest million   6. _____

Name: Date:
Instructor: Section:

Objective 3 Use front end rounding to estimate answers in addition and subtraction.

Video Examples

Review these examples for Objective 3:

6b. Use front end rounding to round 95,602.

The leftmost digit, 9, is in the ten-thousands place, so round to the nearest ten-thousand.
   For 95,602, the next digit is 5 or more.
   Change 9 to 10.
   Regroup 1 into hundred-thousands place.
95,602 rounds to 100,000.

Now Try:

6b. Use front end rounding to round 974,312.

_____

7. Use front end rounding to estimate an answer. Then find the exact answer.
A checkbook had a balance of $912. A check was written for $394. What is the new balance?

Estimate: Use front end rounding to round $912 and $394.
   $912 rounds to $900 $394 rounds to $400
Use the rounded number and subtract to estimate the new checkbook balance.
   $900 − $400 = $500

Exact: Use the original numbers and subtract.
   $912 − $394 = $518

7. Use front end rounding to estimate an answer. Then find the exact answer.
A monthly gross pay is $3207, with deductions of $795. Find the net pay after deductions.

Estimate _____

Exact _____

Objective 3 Practice Exercises

For extra help, see Examples 6–7 on pages 38–39 of your text.

Use front end rounding to round the number.

7. The first printing run of a new book is 170,000 copies.

7. _____

First use front end rounding to estimate each answer. Then find the exact answer.

8. −5172 − 3850

8.
Estimate _____

Exact _____

First use front end rounding to estimate the answer to the application problem. Then find the exact answer.

9. A cannon is raised above the horizontal by an angle of 36°. The cannon is lowered 8°, then lowered another 17°, and finally raised 24°. What is the cannon's final elevation angle?

9.
Estimate _____

Exact _____

Name: Date:
Instructor: Section:

Chapter 1 INTRODUCTION TO ALGEBRA: INTEGERS

1.6 Multiplying Integers

Learning Objectives
1. Used a raised dot or parentheses to express multiplication.
2. Multiply integers.
3. Identify properties of multiplication.
4. Estimate answers to application problems involving multiplication.

Key Terms

Use the vocabulary terms listed below to complete each statement in exercises 1−7.

> **factors**   **product**   **multiplication property of 0**
>
> **multiplication property of 1**   **commutative property of multiplication**
>
> **associative property of multiplication**   **distributive property**

1. By the_____, changing the order of the factors in a multiplication problem does not change the product.

2. In multiplication, the numbers being multiplied are called the _____.

3. By the_____, changing the grouping of the factors in a multiplication problem does not change the product.

4. The _____ says that multiplying a number by 0 gives a product of 0.

5. The answer to a multiplication problem is called the _____.

6. The _____ states that $a(b + c) = ab + bc$.

7. The _____ says that multiplying a number by 1 leaves the number unchanged.

Objective 1   Used a raised dot or parentheses to express multiplication.

Video Examples

Review this example for Objective 1:	Now Try:
1b. Rewrite the multiplication in three different ways, using a dot or parentheses. Also identify the factor and the product. 5×90 Rewrite it as $5 \cdot 90$ or $5(90)$ or $(5)(90)$. The factors are 5 and 90. The product is 450.	**1b.** Rewrite the multiplication in three different ways, using a dot or parentheses. Also identify the factor and the product. 8×15 _____

Name: Date:
Instructor: Section:

Objective 1 Practice Exercises

For extra help, see Example 1 on page 44 of your text.

Rewrite each multiplication, first using a dot and then using parentheses.

1. -3×6

1. _____
   _____

2. $12 \times (-3)$

2. _____
   _____

3. 4×5

3. _____
   _____

Objective 2   Multiply integers.

Video Examples

Review these examples for Objective 2:
Multiply.

2a. $-3 \cdot 9$

The factors have different signs, so the product is negative.
$-3 \cdot 9 = -27$

2b. $-11(-7)$

The factors have the same sign, so the product is positive.
$-11(-7) = 77$

3b. $-3 \cdot (-3) \cdot (-3)$

There is no work to do inside the parentheses, so multiply $-3 \cdot (-3)$ first. The factors have the same sign, so the product is positive.
$-3 \cdot (-3) \cdot (-3)$
$\quad 9 \ \cdot \ (-3)$
$\qquad -27$
Note that the factors 9 and -3 have different signs, so the product is negative.

2c. $8(-12)$

The factors have different signs, so the product is negative.
$8(-12) = -96$

Now Try:
Multiply.

2a. $-6 \cdot 7$

_____

2b. $-4(-15)$

_____

3b. $-4 \cdot (-4) \cdot (-4)$

_____

2c. $9(-6)$

_____

Name: Date:
Instructor: Section:

Objective 2 Practice Exercises

For extra help, see Examples 2–3 on page 46–47 of your text.

Multiply.

4. $-7(-1)(-4)$ 4. _____

5. $-2 \cdot 31$ 5. _____

6. $-5 \cdot (-2) \cdot (6)$ 6. _____

Objective 3   **Identify properties of multiplication.**

Video Examples

Review these examples for Objective 3:

4b. Multiply. Then name the property illustrated by the example.

$763(1)$

$763(1) = 1$

This illustrates the multiplication property of 1.

5. Show that the product is unchanged and name the property that is illustrated in each case.

   a. $-5 \cdot (-8) = -8 \cdot (-5)$

   $-5 \cdot (-8) = -8 \cdot (-5)$
   $ 40 = 40$

   This example illustrates the commutative property of multiplication.

   b. $6 \cdot (7 \cdot 3) = (6 \cdot 7) \cdot 3$

   $6 \cdot (7 \cdot 3) = (6 \cdot 7) \cdot 3$
   $ 6 \cdot 21 = 42 \cdot (3)$
   $ 126 = 126$

   This example illustrates the associative property of multiplication.

Now Try:

4b. Multiply. Then name the property illustrated by the example.
$1(-92)$

_____

5. Show that the product is unchanged and name the property that is illustrated in each case.

   a. $-9 \cdot (-7) = -7 \cdot (-9)$

   _____

   b. $-11 \cdot (3 \cdot 6) = (-11 \cdot 3) \cdot (6)$

   _____

Name: Date:
Instructor: Section:

6b. Rewrite the product using the distributive property. Show that the result is unchanged.

$-3(-8+2)$

$-3(-8+2) = (-3)\cdot(-8)+(-3)\cdot(2)$
$-3(-6) \quad = \quad 24 \quad + \quad (-6)$
$\quad 18 \quad = \quad \quad 18$

Both results are 18.

6b. Rewrite the product using the distributive property. Show that the result is unchanged.
$-9(-5+3)$

_____

Objective 3 Practice Exercises

For extra help, see Examples 4–6 on pages 47–49 of your text.

Fill in the blank to make a true statement.

7. $0 = 18(\_\_\_)$

7. _____

Rewrite each multiplication, using the stated property. Show that the result is unchanged.

8. Commutative property
$-22 \cdot 4$

8. _____

9. Associative property
$(3 \cdot (-2)) \cdot (-9)$

9. _____

Objective 4 Estimate answers to application problems involving multiplication.

Video Examples

Review this example for Objective 4:
7. Use front end rounding to estimate an answer. Then find the exact answer.
A local cable company is losing about 1955 customers per month. How many customers are lost in a year?

Estimate: Use front end rounding: –1955 round to –2000, and 12 months rounds to 10 months.
$(-2000)\cdot(10) = -20,000$ customers

Exact:
$(-1955)\cdot(12) = -23,460$ customers

Now Try:
7. Use front end rounding to estimate an answer. Then find the exact answer.
Enrollment at a community college has decreased by 595 students for the last 3 semesters. What is the total decrease?

Estimate _____

Exact _____

Name: Date:
Instructor: Section:

Objective 4 Practice Exercises

For extra help, see Example 7 on page 49 of your text.

First use front end rounding to estimate the answer to each application problem. Then find the exact answer.

10. An airplane on an approach path to an airport is losing altitude at the rate of 820 feet per minute. How much altitude does the plane lose in 18 minutes?

10.
Estimate_____

Exact _____

11. A television is purchased for $375 down and twelve monthly payments of $55 each. A set of speakers costing a total of $280 is also part of the purchase. What is the total cost of the system?

11.
Estimate_____

Exact _____

12. A tree grows from a seedling for 15 years at a rate of 6 feet per year. At the end of that time, the tree is felled with a cut 4 feet above the ground. Seven feet of the tip are trimmed off. How long is the resulting log?

12.
Estimate_____

Exact _____

Name: Date:
Instructor: Section:

Chapter 1 INTRODUCTION TO ALGEBRA: INTEGERS

1.7 Dividing Integers

Learning Objectives
1. Divide integers.
2. Identify properties of division.
3. Combine multiplying and dividing of integers.
4. Estimate answers to application problems involving division.
5. Interpret remainders in division application problems.

Key Terms

Use the vocabulary terms listed below to complete each statement in exercises 1−3.

    factors   product   quotient

1. Numbers that are being multiplied are called _____.

2. The answer to a division problem is called the _____.

3. The answer to a multiplication problem is called the _____.

Objective 1   Divide integers.

Video Examples

Review these examples for Objective 1:
1. Divide.

   a. $\dfrac{-30}{6}$

   The numbers have different signs, so the quotient is negative.
   $$\dfrac{-30}{6} = -5$$

   b. $\dfrac{-27}{-9}$

   The numbers have the same sign, so the quotient is positive.
   $$\dfrac{-27}{-9} = 3$$

   c. $70 \div (-5)$

   The numbers have different signs, so the quotient is negative.
   $$70 \div (-5) = -14$$

Now Try:
1. Divide.

   a. $\dfrac{-35}{7}$

   _____

   b. $\dfrac{-40}{-10}$

   _____

   c. $24 \div (-8)$

   _____

Name: Date:
Instructor: Section:

Objective 1 Practice Exercises

For extra help, see Example 1 on page 54 of your text.

Divide.

1. $\dfrac{28}{-4}$ 1. _____

2. $\dfrac{-36}{-12}$ 2. _____

3. $\dfrac{-100}{10}$ 3. _____

Objective 2   Identify properties of division.

Video Examples

Review these examples for Objective 2:

2. Divide. Then state the property illustrated by each example.

   a. $\dfrac{-672}{-672}$

   Any nonzero number divided by itself is 1.
   $\dfrac{-672}{-672} = 1$

   b. $\dfrac{85}{1}$

   Any number divided by 1 is the number.
   $\dfrac{85}{1} = 85$

   c. $\dfrac{0}{23}$

   Zero divided by any nonzero number is 0.
   $\dfrac{0}{23} = 0$

   d. $\dfrac{67}{0}$

   Division by 0 is undefined.
   $\dfrac{67}{0}$ is undefined.

Now Try:

2. Divide. Then state the property illustrated by each example.

   a. $\dfrac{-703}{-703}$

   _____

   b. $\dfrac{93}{1}$

   _____

   c. $\dfrac{0}{39}$

   _____

   d. $\dfrac{51}{0}$

   _____

Name: Date:
Instructor: Section:

Objective 2 Practice Exercises

For extra help, see Example 2 on page 55 of your text.

Complete the equation, then state the property represented by the equation.

4. $\dfrac{0}{14} =$ 4. _____

5. $\dfrac{-6}{-6} =$ 5. _____

6. $\dfrac{-48}{0} =$ 6. _____

Objective 3   Combine multiplying and dividing of integers.

Video Examples

Review these examples for Objective 3:	**Now Try:**
3. Simplify.	3. Simplify.
c. $-80 \div (-16) \div (-1)$	**c.** $-56 \div (-8) \div (-7)$
$-80 \div (-16) \div (-1)$	_____
$5 \div (-1)$	
-5	
a. $8(-12) \div (-4 \cdot 2)$	**a.** $5(-16) \div (-2 \cdot 4)$
$8(-12) \div (-4 \cdot 2)$	_____
$8(-12) \div (-8)$	
$-96 \div (-8)$	
12	

Objective 3 Practice Exercises

For extra help, see Example 3 on page 56 of your text.

Simplify.

7. $-120 \div (-6) \div 10$ 7. _____

Name: Date:
Instructor: Section:

8. $28 \cdot (2 \cdot (-1)) \div -8$

8. _____

9. $84 \div (-3) \cdot 3 \div 4$

9. _____

Objective 4   Estimate answers to application problems involving division.

Video Examples

Review this example for Objective 4:

4. Use front end rounding to estimate an answer. Then find the exact answer.
Matthew lost 108 pounds last year. What was the average weight loss per month?

Estimate: Use front end rounding: −108 rounds to −100, and 12 months rounds to 10 months.
    $-100 \div 10 = -10$ pounds each month

Exact:
    $-108 \div 12 = -9$ pounds each month

Now Try:

4. Use front end rounding to estimate an answer. Then find the exact answer.
During the last year, John lost $3708 in the stock market. What was the average loss each month?

_____

Objective 4 Practice Exercises

For extra help, see Example 4 on page 57 of your text.

Solve these application problems by using addition, subtraction, multiplication, or division. First estimate the answer using front end rounding; then find the exact answer.

10. A honeybee flies from its hive a distance of 52 meters to one flowering bush, and from there it flies a distance of 17 meters from there to the next bush. What is the bee's total distance of travel so far?

10.
Estimate _____

Exact _____

Name: Date:
Instructor: Section:

11. Lila has a 36-month car payment of $22,500. How much is her monthly payment?

11.
Estimate_____

Exact _____

12. A sound wave has a frequency of 468 cycles per second. This frequency is multiplied by 5 to produce a new sound. What is the new frequency?

12.
Estimate_____

Exact _____

Objective 5 Interpreting Remainders in Division Applications.

Video Examples

Review this example for Objective 5:

5b. Divide; then interpret the remainder in the application.

The college marching band has 118 members and will spend the night in a hotel. Each room can accommodate 4 people. How many rooms are needed?

We use division to solve the problem.

```
      29
   ┌─────
 4 )118
      8
     ──
     38
     36
     ──
      2
```

If 29 rooms are rented, two members will have to sleep on a bus. So, 30 rooms must be rented. (One room will have only 2 band members.)

Now Try:

5b. Divide; then interpret the remainder in the application.

In the lobby of a skyscraper, 150 people are waiting for elevators. If the capacity of each elevator is 18 people, how many elevator trips are needed?

_____

Name: Date:
Instructor: Section:

Objective 5 Practice Exercises

For extra help, see Example 5 on pages 57–58 of your text.

Divide, then interpret the remainder in each application.

13. A group of 96 construction workers is to be transported on 7 trucks. The workers are to be divided as evenly as possible. How many workers will be transported on each truck?

13. _____

14. A student plans to memorize 213 pages of notes in seven days. How many pages per day should the student memorize?

14. _____

15. A graduation ceremony will be held in an auditorium with 1500 seats. If there are 468 students in the class, how many tickets will be allotted to each student?

15. _____

Name: Date:
Instructor: Section:

Chapter 1 INTRODUCTION TO ALGEBRA: INTEGERS

1.8   Exponents and Order of Operations

Learning Objectives
1   Use exponents to write repeated factors.
2   Simplify expressions containing exponents.
3   Use the order of operations.
4   Simplify expressions with fraction bars.

Key Terms

Use the vocabulary terms listed below to complete each statement in exercises 1–2.

    exponent   **order of operations**

1. For problems or expressions with more than one operation, the _____ tells what to do first, second, and so on, to obtain the correct answer.

2. An _____ tells how many times a number is used as a factor in repeated multiplication.

Objective 1   Use exponents to write repeated factors.

Video Examples

Review these examples for Objective 1:	**Now Try:**

1. Given the factored form, give the exponential form, simplified form, and how it is read.
 	
    a. $4 \cdot 4 \cdot 4$
 	
    Exponential form: 4^3
    Simplified: 64
    Read as: 4 cubed, or 4 to the third power
 	
    b. $(3)(3)$
 	
    Exponential form: 3^2
    Simplified: 9
    Read as: 3 squared, or 3 to the second power
 	
    c.   11
 	
    Exponential form: 11^1
    Simplified: 11
    Read as: 11 to the first power
	1. Given the factored form, give the exponential form, simplified form, and how it is read.
	    a. $5 \cdot 5 \cdot 5 \cdot 5$
	
	    _____
	
	    b. $(6)(6)$
	
	    _____
	
	    c. 23
	
	    _____

Name: Date:
Instructor: Section:

Objective 1 Practice Exercises

For extra help, see Example 1 on page 67 of your text.

Rewrite each number in factored form as a number in exponential form; then state how the exponential form is read.

1. $12 \cdot 12 \cdot 12$

   1. _____
      _____

2. $6 \cdot 6 \cdot 6 \cdot 6 \cdot 6$

   2. _____
      _____

3. $7 \cdot 7 \cdot 7 \cdot 7 \cdot 7 \cdot 7 \cdot 7$

   3. _____
      _____

Objective 2 Simplify expressions containing exponents.

Video Examples

Review these examples for Objective 2:
2. Simplify.

   a. $(-4)^2$

   $(-4)^2 = (-4)(-4) = 16$

   c. $(-4)^4$

   $(-4)^4 = (-4)(-4)(-4)(-4) = 256$

   b. $(-3)^3$

   $(-3)^3 = (-3)(-3)(-3)$
   $ = 9(-3)$
   $ = -27$

Now Try:
2. Simplify.

   a. $(-7)^2$

   _____

   c. $(-3)^4$

   _____

   b. $(-10)^3$

   _____

Name: Date:
Instructor: Section:

Objective 2 Practice Exercises

For extra help, see Example 2 on page 68 of your text.

Simplify.

4. $(-1)^{91}$ 4. _____

5. $2^4 \cdot (-3)^2$ 5. _____

6. $(-2)^5 \cdot (-3)^2$ 6. _____

Objective 3 Use the order of operations.

Video Examples

Review these examples for Objective 3:	Now Try:
6a. Simplify.	6a. Simplify.
$6^2 - (-7)^2$	$8^2 - (-9)^2$
$6^2 - (-7)^2$	
$6^2 - 49$	_____
$36 - 49$	
-13	
4. Simplify $8 + 5(24 - 6) \div 9$.	4. Simplify $5 + 4(25 - 3) \div 11$.
$8 + 5(24 - 6) \div 9$	
$8 + 5\ (18)\ \div 9$	
$8 + 90 \div 9$	_____
$8 + 10$	
18	

Name: Date:
Instructor: Section:

6b. Simplify.

$(-3)^4 - (5-7)^2(-9)$

$(-3)^4 - (5-7)^2(-9)$

$(-3)^4 - (-2)^2(-9)$

$81 - 4(-9)$

$81 - (-36)$

$81 + 36$

117

5a. Simplify.

$-9 \div (7-4) - 13$

$-9 \div (7-4) - 13$

$-9 \div (3) - 13$

$-3 - 13$

-16

6b. Simplify.

$(-5)^3 - (6-9)^2(-8)$

_____

5a. Simplify.

$-10 \div (8-6) - 19$

_____

Objective 3 Practice Exercises

For extra help, see Examples 3–6 on pages 68–71 of your text.

Simplify.

7. $3(-2+6) - (8-13)$

7. _____

8. $3 - (-6) \cdot (-1)^8$

8. _____

9. $(-2)^3 \cdot (7-9)^2 \div 2$

9. _____

Name: Date:
Instructor: Section:

Objective 4 Simplify expressions with fraction bars.

Video Examples

Review this example for Objective 4:

7. Simplify $\dfrac{-9+6(7-10)}{5-6^2 \div 18}$.

First do work in the numerator.
$-9+6(7-10)$
$-9+\ 6(-3)$
$-9+(-18)$
$\quad -27 \leftarrow$ Numerator

Now do the work in the denominator.
$5-6^2 \div 18$
$5-36 \div 18$
$5-2$
$\quad 3 \leftarrow$ Denominator

The last step is the division.
$\begin{array}{l}\text{Numerator} \to \\ \text{Denominator} \to\end{array} \dfrac{-27}{3} = -9$

Now Try:

7. Simplify $\dfrac{-15+7(9-12)}{8-10^2 \div 25}$.

_____

Objective 4 Practice Exercises

For extra help, see Example 7 on page 72 of your text.

Simplify.

10. $\dfrac{-8+4^2-(-9)}{11-3-9}$

10. _____

11. $\dfrac{-3 \cdot 4^2 - 2(5+(-5))}{-3(8-13) \div -5}$

11. _____

12. $\dfrac{4 \cdot 2^2 - 8(7-2)}{7(6-8) \div 14}$

12. _____

Name: Date:
Instructor: Section:

Chapter 2 UNDERSTANDING VARIABLES AND SOLVING EQUATIONS

2.1 Introduction to Variables

Learning Objectives
1. Identify variables, constants, and expressions.
2. Evaluate variable expressions for given replacement values.
3. Write properties of operations using variables.
4. Use exponents with variables.

Key Terms

Use the vocabulary terms listed below to complete each statement in exercises 1−5.

variable   constant   expression

evaluate the expression   coefficient

1. An _____ tells the rule for doing something.

2. A _____ is a letter that represents a number that varies or changes, depending on the situation.

3. To _____, replace each variable with specific values and then follow the order of operations.

4. A _____ is a number that is added or subtracted in an expression.

5. The number part in a multiplication expression is the _____.

Objective 1 Identify variables, constants, and expressions.

Video Examples

Review this example for Objective 1:
1. Write an expression for this rule. Identify the variable and the constant.
   Order the class limit minus 8 lunches because some students will brown bag.

   Let c represent the variable and 8 is the constant.
   $c - 8$

Now Try:
1. Write an expression for this rule. Identify the variable and the constant.
   Order the class limit plus 5 extra lunches because some students are football players.

   _____

Name: Date:
Instructor: Section:

Objective 1 Practice Exercises

For extra help, see Example 1 on page 95 of your text.

Identify the parts of each expression. Choose from **variable**, **constant**, *and* **coefficient**.

1. $-7 + h$ 1. _____

                                         _____

2. $-2w$ 2. _____

                                         _____

3. $9k + 1$ 3. _____

                                         _____

Objective 2   Evaluate variable expressions for given replacement values.

Video Examples

Review these examples for Objective 2:

2. Use this rule (from Example 1) for ordering lunches: Order the class limit minus 8. The expression is $c - 8$.

   a. Evaluate the expression when the class limit is 39.

   Replace c with 39 and follow the rule.
   $c - 8$
   $39 - 8$
   31   Order 31 lunches.

   b. Evaluate the expression when the class limit is 22.

   Replace c with 22 and follow the rule.
   $c - 8$
   $22 - 8$
   14   Order 14 lunches.

Now Try:

2. Use this rule (from Now Try 1) for ordering lunches: Order the class limit plus 8. The expression is $c + 5$.

   a. Evaluate the expression when the class limit is 24.

   _____

   b. Evaluate the expression when the class limit is 45.

   _____

Name: Date:
Instructor: Section:

3. The expression (rule) for finding the perimeter of a square shape is 4*s*. Evaluate the expression when the length of a side of a square garden is 20 feet.

   Replace *s* with 20 feet.

   The total distance around the square garden is 80 feet.

5a. Find your average score if you bowl three games and your total score for all three games is 384.

   Use the expression (rule) for finding your average score. Replace *t* with your total score of 384, and replace *g* with 3, the number of games.

   $\frac{t \rightarrow 384}{g \rightarrow 3}$

   Divide 384 by 3 to get 128. Your average score is 128.

3. The expression (rule) for finding the perimeter of a square shape is 4*s*. Evaluate the expression when the length of a side of a square garden is 8 yards.

   _____

5a. Find your average score if you bowl four games and your total score for all four games is 384.

   _____

Objective 2 Practice Exercises

For extra help, see Examples 2–5 on pages 95–98 of your text.

Evaluate each expression.

4. The expression (rule) for the weight of an object (in Newtons) is 10*m*, where *m* is the object's mass in kilograms. Evaluate the expression when
   (a) the object has a mass of 14 kilograms.
   (b) the object has a mass of 92 kilograms.

   4. a._____
      b._____

5. The expression (rule) for the total number of dollars spent on a car, when there is a $2300 down payment and monthly payments of $215, is $2300 + 215*t*, where *t* is the number of monthly payments. Evaluate the expression when
   (a) there are 36 monthly payments.
   (b) there are 48 monthly payments.

   5. a._____
      b._____

Name: Date:
Instructor: Section:

6. The expression (rule) for the amount of insurance paid out on a policy with a $2500 deductible is $\frac{8c}{10} - \$2500$, where c is total medical costs. Evaluate the expression when
   (a) total medical costs are $10,625.
   (b) total medical costs are $12,480.

6. a._____

   b._____

Objective 3 Write properties of operations using variables.

Video Examples

Review this example for Objective 3:
6. Use the variable b to state the property: When any number is added to 0, the sum is the number.

   Use the letter b to represent any number.
   $b + 0 = b$

Now Try:
6. Use the variable b to state the property: When any non-zero number is divided into 0, the quotient is zero.

   _____

Objective 3 Practice Exercises

For extra help, see Example 6 on page 99 of your text.

State the property represented by each equation.

7. $h(a + b) = h(a) + h(b)$

7. _____

8. $(1)(k) = k$

8. _____

9. $x + 0 = x$

9. _____

Name: Date:
Instructor: Section:

Objective 4   Use exponents with variables.

Video Examples

Review these examples for Objective 4:

7. Rewrite each expression without exponents.

   a. x^6

   x^6 can be written as $x \cdot x \cdot x \cdot x \cdot x \cdot x$.

   b. $13a^2b$

   $13a^2b$ can be written as $13 \cdot a \cdot a \cdot b$.

   c. $-10p^3q^4$

   $-10p^3q^4$ can be written as $-10 \cdot p \cdot p \cdot p \cdot q \cdot q \cdot q \cdot q$.

8. Evaluate each expression.

   b. x^2y^3 when x is -6 and y is -3

   Replace x with -6 and replace y with -3.
x^2y^3 means $x \cdot x \cdot y \cdot y \cdot y$
                   $(-6)\cdot(-6)\cdot(-3)\cdot(-3)\cdot(-3)$
                       $36 \quad \cdot(-3)\cdot(-3)\cdot(-3)$
                       $-108 \quad \cdot(-3)\cdot(-3)$
                           $324 \quad \cdot(-3)$
                                  -972

So x^2y^3 becomes $(-6)^2(-3)^3$, which is $(-6)\cdot(-6)\cdot(-3)\cdot(-3)\cdot(-3)$, or -972.

   c. $-7a^2b$ when a is 4 and b is 6

   Replace a with 4 and replace b with 6.
$-7a^2b$ means $-7 \cdot a \cdot a \cdot b$
                     $-7 \cdot 4 \cdot 4 \cdot 6$
                     $-28 \cdot 4 \cdot 6$
                     $-112 \cdot 6$
                     -672

So $-7a^2b$ becomes $-7(4)^2(6)$, which is $-7 \cdot 4 \cdot 4 \cdot 6$, or -672.

Now Try:

7. Rewrite each expression without exponents.

   a. z^4

   _____

   b. $14rs^2$

   _____

   c. $-13m^3n^5$

   _____

8. Evaluate each expression.

   b. x^4y^2 when x is -2 and y is -3

   _____

   c. $-6c^3d$ when c is 2 and d is 5

   _____

Name: Date:
Instructor: Section:

Objective 4 Practice Exercises

For extra help, see Examples 7–8 on page 100 of your text.

Rewrite each expression without exponents.

10. $-7g^3h$

10. _____

11. $u^3v^2w^2$

11. _____

Evaluate the expression.

12. $-3k^2g^3$ when k is -2 and g is 5

12. _____

Name: Date:
Instructor: Section:

Chapter 2 UNDERSTANDING VARIABLES AND SOLVING EQUATIONS

2.2 Simplifying Expressions

Learning Objectives
1 Combine like terms using the distributive property.
2 Simplify expressions.
3 Use the distributive property to multiply.

Key Terms

Use the vocabulary terms listed below to complete each statement in exercises 1–5.

simplify an expression **term** **constant term**
variable term **like terms**

1. _____ are terms with exactly the same variable parts.

2. Each addend in an expression is called a _____.

3. To _____, combine all the like terms.

4. A _____ has a coefficient multiplied by a variable part.

5. A _____ is a type of term that is just a number.

Objective 1   Combine like terms using the distributive property.

Video Examples

Review this example for Objective 1:
2a. Combine like terms.

$7x + x + 9x$

$7x + x + 9x$

$7x + 1x + 9x$

$(7 + 1 + 9)x$

$17x$

Now Try:
2a. Combine like terms.

$12x + 7x + x$

_____

Objective 1 Practice Exercises

For extra help, see Examples 1–2 on pages 108–109 of your text.

Identify the like terms in the expression. Then identify the coefficients of the like terms.

1. $x^2 + 2x + (-5x^3) + (-3x) + 8$

1. _____

   _____

Copyright © 2018 Pearson Education, Inc.   39

Name: Date:
Instructor: Section:

Simplify each expression.

2. $6d + d$

2. _____

3. $c^3z^4 - 11c^3z^4 - 9c^3z^4$

3. _____

Objective 2   Simplify expressions.

Video Examples

Review these examples for Objective 2:

3b. Simplify the expression by combining like terms.

$4c - 9 - c + 11$

Write 1 as the coefficient of c. Change subtractions to adding the opposite, and then combine like terms.

$4c - 9 - 1c + 11$
$4c + (-9) + (-1c) + 11$
$4c + (-1c) + (-9) + 11$
$[4 + (-1)]c + (-9) + 11$
$\quad 3c \quad + \quad 2$

The simplified expression is $3c + 2$.

4c. Simplify.

$-9(-3a^2)$

Use the associative property.

$-9(-3a^2)$ can be written as $[-9 \cdot (-3)]a^2$
$\qquad\qquad 27a^2$

So, $-9(-3a^2)$ simplifies to $27a^2$.

Now Try:

3b. Simplify the expression by combining like terms.
$8c - 13 - c + 4$

_____

4c. Simplify.

$-8(-5y^2)$

_____

Name: Date:
Instructor: Section:

Objective 2 Practice Exercises

For extra help, see Examples 3–4 on pages 110–111 of your text.

Simplify each expression by combining like terms. Write each answer with the variables in alphabetical order and any constant term last.

4. $5b^4m + 10 - 2b^4m - 1$ 4. _____

5. $-8s + 6st - 9t - 11st + 2s + 4t + 9 - 11t - 1$ 5. _____

Simplify by using the associative property of multiplication.

6. $-14(-a^3c^2t)$ 6. _____

Objective 3   Use the distributive property to multiply.

Video Examples

Review these examples for Objective 3:
5. Simplify.

   c. $8(d-9)$

   $8(d-9)$ can be written as $8[d+(-9)]$
   $$8 \cdot d + 8(-9)$$
   $$8d + (-72)$$
   $$8d - 72$$
   So, $8(d-9)$ simplifies as $8d - 72$.

   a. $7(4x+3)$

   $7(4x+3)$ can be written as $7 \cdot 4x + 7 \cdot 3$
   $$7 \cdot 4 \cdot x + 21$$
   $$28 \cdot x + 21$$
   $$28x + 21$$
   So, $7(4x+3)$ simplifies as $28x + 21$.

Now Try:
5. Simplify.

   c. $28(x-4)$

   _____

   a. $6(5x+2)$

   _____

Name: Date:
Instructor: Section:

b. $-5(8b+4)$

$-5(8b+4)$ can be written as $-5 \cdot 8b + (-5) \cdot 4$
$-5 \cdot 8 \cdot b + (-20)$
$-40 \cdot b + (-20)$
$-40b + (-20)$

So, $-5(8b+4)$ simplifies as $-40b - 20$.

6. Simplify: $9 + 4(x-5)$

Use the distributive property first.
$9 + 4(x-5)$
$9 + 4 \cdot x - 4 \cdot 5$
$9 + 4x - 20$
$4x + 9 - 20$
$4x\ -11$

The simplified expression is $4x - 11$.

b. $-9(6x+4)$

_____

6. Simplify: $12 + 6(x-3)$

_____

Objective 3 Practice Exercises

For extra help, see Examples 5–6 on pages 112–113 of your text.

Use the distributive property to simplify each expression.

7. $-5(t+4)$ **7.** _____

8. $-8(3s-4)$ **8.** _____

Simplify the expression.

9. $-q + 7(3q-2) + 11$ **9.** _____

42   Copyright © 2018 Pearson Education, Inc.

Name: Date:
Instructor: Section:

Chapter 2 UNDERSTANDING VARIABLES AND SOLVING EQUATIONS

2.3 Solving Equations Using Addition

Learning Objectives
1 Determine whether a given number is a solution of an equation.
2 Solve equations using the addition property of equality.
3 Simplify equations before using the addition property of equality.

Key Terms

Use the vocabulary terms listed below to complete each statement in exercises 1–5.

    equation   **solve an equation**   **solution**

    addition property of equality   **check the solution**

1. A _____ of an equation is a number that makes the statement true when it is substituted for the variable.

2. The _____ states that adding the same quantity to both sides of an equation keeps the equation balanced.

3. An _____ is a mathematical statement that contains an equals sign.

4. To _____, find a number that can replace the variable and make the equation balance.

5. To _____, go back to the original equation and replace the variable with the solution.

Objective 1 Determine whether a given number is a solution of an equation.

Video Examples

Review this example for Objective 1:
1. Which of these numbers, 65, 45, or 85, is the solution of the equation $c - 15 = 70$?

   Replace c with each of the numbers. The one that makes the equation balance is the solution.
   $65 - 15 \neq 70$ Does not balance: $65 - 15 = 50$.

   $45 - 15 \neq 70$ Does not balance: $45 - 15 = 30$.

   $85 - 15 = 70$ Balances: $85 - 15 = 70$.

   The solution is 85 because when c is 85, the equation balances.

Now Try:
1. Which of these numbers, 25, 29, or 21, is the solution of the equation $25 = c - 4$?

   _____

Name: Date:
Instructor: Section:

Objective 1 Practice Exercises

For extra help, see Example 1 on page 123 of your text.

In each list of numbers, find the one that is a solution of the given equation.

1. $r - 12 = 0$
   $-8, -2, 12$

   1. _____

2. $p + 7 = 10$
   $-7, 3, 7$

   2. _____

3. $6 - 2y = -8$
   $-3, 5, 7$

   3. _____

Objective 2 Solve equations using the addition property of equality.

Video Examples

Review these examples for Objective 2:

2. Solve each equation and check the solution.

   b. $-7 = x - 4$

   Change subtraction to adding the opposite. Then Add the opposite of –4, which is 4.
   $-7 = x - 4$
   $-7 = x + (-4)$
   $\underline{4 \quad 4}$
   $-3 = x + 0$
   $-3 = x$

   Because x balances with –3, the solution is –3.

Now Try:
2. Solve each equation and check the solution.
   b. $-10 = x - 8$

   _____

Name: Date:
Instructor: Section:

Check the solution by replacing c with -3 in the original equation.
$$-7 = x - 4$$
$$-7 = -3 - 4$$
$$-7 = -3 + (-4)$$
$$-7 = -7$$

When x is replaced with -3, the equation balances, so -3 is the correct solution.

a. $c + 7 = 20$

a. $c + 13 = 29$

Add -7 to each side.
$$c + 7 = 20$$
$$\underline{-7 \quad -7}$$
$$c + 0 = 13$$
$$c = 13$$

Because c balances with 13, the solution is 13.

Check the solution by replacing c with 13 in the original equation.
$$c + 7 = 20$$
$$13 + 7 = 20$$
$$20 = 20$$

The equation balances. Therefore, 13 is the correct solution.

Objective 2 Practice Exercises

For extra help, see Example 2 on pages 125–126 of your text.

Solve each equation and check the solution.

4. $v - 72 = 32$

4. _____

5. $-30 = 36 + n$

5. _____

Name: Date:
Instructor: Section:

6. $65 = u + 74$ **6.** _____

Objective 3 Simplify equations before using the addition property of equality.

Video Examples

Review these examples for Objective 3:
3. Solve each equation and check the solution.

 a. $y + 9 = 5 - 8$

Simplify the right side first. Then get y by itself on the left side by adding the opposite of 9, which is -9.

$y + 9 = 5 - 8$
$y + 9 = 5 + (-8)$
$y + 9 = -3$
$\underline{-9 \quad -9}$
$y = -12$

The solution is -12. Now check the solution.

Check Go back to the original equation and replace y with -12.
$y + 9 = 5 - 8$
$-12 + 9 = 5 - 8$
$-3 = 5 + (-8)$
$-3 = -3$

So when y is replaced with -12, the equation balances, so -12 is the correct solution.

 b. $-5 + 5 = -7b - 10 + 8b$

Combine like terms on each side first.
$-5 + 5 = -7b - 10 + 8b$
$0 = -7b + (-10) + 8b$
$0 = -7b + 8b + (-10)$
$0 = 1b + (-10)$
$\underline{10 10}$
$10 = b$

The solution is 10.

Now Try:
3. Solve each equation and check the solution.

 a. $x + 10 = 14 - 15$

 _____

 b. $-9 + 9 = -15b - 6 + 16b$

 _____

Name: Date:
Instructor: Section:

Check Go back to the original equation and replace b with 10.
$$-5+5=-7b-10+8b$$
$$-5+5=-7\cdot 10+(-10)+8\cdot 10$$
$$0 = -70+(-10)+80$$
$$0 = -80 +80$$
$$0 = 0$$
When b is replaced with 10, the equation balances, so 10 is the correct solution.

Objective 3 Practice Exercises

For extra help, see Example 3 on page 126–127 of your text.

Simplify each side of the equation when possible. Then solve the equation. Check each solution.

7. $11a - 10a = -5 + 8$

7. _____

8. $40 - 2u + 3u = 19$

8. _____

9. $66n + 53 - 65n = 9$

9. _____

Name: Date:
Instructor: Section:

Chapter 2 UNDERSTANDING VARIABLES AND SOLVING EQUATIONS

2.4 Solving Equations Using Division

Learning Objectives
1. Solve equations using the division property of equality.
2. Simplify equations before using the division property of equality.
3. Solve equations such as $-x = 5$.

Key Terms

Use the vocabulary terms listed below to complete each statement in exercises 1–2.

division property of equality

addition property of equality

1. The _____ states that dividing both sides of an equation by the same nonzero number will keep it balanced.

2. When the same quantity is added to both sides of an equation, the _____ is being applied.

Objective 1 Solve equations using the division property of equality.

Video Examples

Review this example for Objective 1:
1b. Solve the equation and check the solution.

$$56 = -7w$$

On the right side of the equation, the variable is multiplied by –7. To undo the multiplication, divide by –7.

$$56 = -7w$$

$$\frac{56}{-7} = \frac{-7w}{-7}$$

$$-8 = w$$

The solution is –8.

Check
We check the solution in the original equation.

$$56 = -7w$$

$$56 = -7 \cdot (-8)$$

$$56 = 56$$

When w is replaced with –8, the equation balances, so –8 is the correct solution.

Now Try:
1b. Solve the equation and check the solution.
$$64 = -4w$$

_____

Name: Date:
Instructor: Section:

Objective 1 Practice Exercises

For extra help, see Example 1 on page 135 of your text.

Solve each equation and check.

1. $-3m = 21$

   1. _____

2. $-48 = -6b$

   2. _____

3. $15h = 420$

   3. _____

Objective 2   Simplify equations before using the division property of equality.

Video Examples

Review these examples for Objective 2:

2. Solve each equation and check each solution.

   a. $6y - 10y = -20$

   First, simplify the left side by combining like terms.
   $$6y - 10y = -20$$
   $$6y + (-10y) = -20$$
   $$\frac{-4y}{-4} = \frac{-20}{-4}$$
   $$y = 5$$

   The solution is 5.

   Check
   Go back to the original equation and replace each *y* with 5.

Now Try:

2. Solve each equation and check each solution.

   a. $5y - 14y = -36$

   _____

Name: Date:
Instructor: Section:

$$6y - 10y = -20$$
$$6 \cdot 5 - 10 \cdot 5 = -20$$
$$30 - 50 = -20$$
$$30 + (-50) = -20$$
$$-20 = -20$$

When y is replaced with 5, the equation balances, so 5 is the correct solution.

b. $13 - 21 + 8 = h + 6h$

First, combine like terms.
$$13 - 21 - 8 = h + 6h$$
$$13 + (-21) - 8 = 1h + 6h$$
$$-8 \quad -8 = 7h$$
$$\frac{0}{7} = \frac{7h}{7}$$
$$0 = h$$

The solution is 0.

Check Go back to the original equation and replace each h with 0.
$$13 - 21 - 8 = h + 6h$$
$$13 + (-21) - 8 = 0 + 6 \cdot 0$$
$$-8 \quad -8 = 0 + 0$$
$$0 = 0$$

When h is replaced with 0, the equation balances, so 0 is the correct solution.

b. $24 - 43 + 19 = h - 9h$

b. _____

Objective 2 Practice Exercises

For extra help, see Example 2 on page 136 of your text.

Solve each equation and check.

4. $53 - 8 = -9d$ 4. _____

Name: Date:
Instructor: Section:

5. $35m - 33m = 100 - 34$ 5. _____

6. $140 - 116 = 20x - 12x$ 6. _____

Objective 3 Solve equations such as $-x = 5$.

Video Examples

Review this example for Objective 3:

3. Solve $-x = 9$ and check the solution.

Start by writing in the coefficient of x.
$$-x = 9$$
$$-1x = 9$$
$$\frac{-1x}{-1} = \frac{9}{-1}$$
$$x = -9$$
The solution is –9.

Check Go back to the original equation and replace x with 0.
$$-x = 9$$
$$-(-9) = 9$$
$$9 = 9$$
When x is replaced with –9, the equation balances, so –9 is the correct solution.

Now Try:

3. Solve $-y = 13$ and check the solution.

_____

Name: Date:
Instructor: Section:

Objective 3 Practice Exercises

For extra help, see Example 3 on page 137 of your text.

Solve each equation and check.

7. $-b = 19$ 7. _____

8. $-17 = -w$ 8. _____

9. $4 = -v$ 9. _____

Name: Date:
Instructor: Section:

Chapter 2 UNDERSTANDING VARIABLES AND SOLVING EQUATIONS

2.5 Solving Equations with Several Steps

Learning Objectives
1. Solve equations using the addition and division properties of equality.
2. Solve equations using the distributive, addition, and division properties.

Key Terms

Use the vocabulary terms listed below to complete each statement in exercises 1–3.

distributive property

division property of equality

addition property of equality

1. The _____ states that if $a = b$, then $a + c = b + c$.

2. The _____ states that $a(b + c) = ab + bc$.

3. The _____ states that if $a = b$, then $\dfrac{a}{c} = \dfrac{b}{c}$ for $c \neq 0$.

Objective 1 Solve equations using the addition and division properties of equality.

Video Examples

Review these examples for Objective 1:

1. Solve this equation and check the solution: $7m + 2 = 37$.

   Step 1 Adding -2 to the left side will leave $7m$ by itself. To keep the balance, add -2 to the right side also.

   $7m + 2 = 37$
   $\underline{-2 \quad -2}$
   $7m + 0 = 35$
   $7m = 35$

   Step 2 Divide both sides by the coefficient of the variable term, 7.

   $\dfrac{7m}{7} = \dfrac{35}{7}$
   $m = 5$

Now Try:

1. Solve this equation and check the solution: $8n + 3 = 59$.

   _____

Name: Date:
Instructor: Section:

Step 3 Check the solution by going back to the original equation. Replace *m* with 5.

$$7m + 2 = 37$$
$$7(5) + 2 = 37$$
$$35 + 2 = 37$$
$$37 = 37$$

When *m* is replaced with 5, the equation balances, so 5 is the correct solution.

2. Solve this equation and check the solution: $5x - 6 = 8x - 18$.

To keep the variable on the left side, add the opposite of $8x$, which is $-8x$ to both sides.

$$\begin{aligned} 5x - 6 &= 8x - 18 \\ \underline{-8x} & \underline{-8x} \\ -3x - 6 &= 0 - 18 \\ -3x + (-6) &= 0 + (-18) \\ \underline{6} & \underline{6} \\ -3x + 0 &= 0 + (-12) \\ \frac{-3x}{-3} &= \frac{-12}{-3} \\ x &= 4 \end{aligned}$$

Check
$$5x - 6 = 8x - 18$$
$$5(4) - 6 = 8(4) - 18$$
$$20 + (-6) = 32 + (-18)$$
$$14 = 14$$

When *x* is replaced with 4, the equation balances, so 4 is the correct solution.

2. Solve this equation and check the solution: $8x - 15 = 10x + 5$.

2. _____

Objective 1 Practice Exercises

For extra help, see Examples 1–2 on pages 142–143 of your text.

Solve each equation and then check.

1. $4m - 19 = -19$

1. _____

Name: Date:
Instructor: Section:

2. $-14+8d = -24+3d$ 2. _____

3. $5q + 10 = 19 + 8q$ 3. _____

Objective 2   Solve equations using the distributive, addition, and division properties.

Video Examples

Review these examples for Objective 2:
3. Solve this equation and check the solution: $-12 = 6(y-2)$.

   We can use the distributive property to simplify the right side of the equation. Then use the steps to solve for y.
   $$-12 = 6(y-2)$$
   $$-12 = 6 \cdot y - 6 \cdot 2$$
   $$-12 = 6y - 12$$
   $$-12 = 6y + (-12)$$
   $$\underline{12 \underline{12}}$$
   $$0 = 6y$$
   $$\frac{0}{6} = \frac{6y}{6}$$
   $$0 = y$$
   The solution is 0.

Now Try:
3. Solve this equation and check the solution: $8(m-3) = -24$.

   _____

Name: Date:
Instructor: Section:

Check Go back to the original equation and replace y with 0.
$$-12 = 6(y-2)$$
$$-12 = 6(0-2)$$
$$-12 = 6[0+(-2)]$$
$$-12 = 6(-2)$$
$$-12 = -12$$
When y is replaced with 0, the equation balances, so 0 is the correct solution.

4. Solve this equation and check the solution: $3+6(n+5) = 5+2n$.

   Step 1 Use the distributive property on the left side.
   $$3+6(n+5) = 5+2n$$
   $$3+6n+30 = 5+2n$$

   Step 2 Combine like terms on the left side.
   $$6n+33 = 5+2n$$

   Step 3 Add $-2n$ to both sides.
   $$\underline{-2n} \quad \underline{-2n}$$
   $$4n+33 = 5+0$$
   $$4n+33 = 5$$

   Step 3 To get $4n$ by itself, add -33 to both sides.
   $$\underline{-33} \quad \underline{-33}$$
   $$4n+0 = -28$$

   Step 4 Divide both sides by 4, the coefficient of the variable term $4n$.
   $$\frac{4n}{4} = \frac{-28}{4}$$
   $$n = -7$$

   Step 5 Check
   $$3+6(n+5) = 5+2n$$
   $$3+6(-7+5) = 5+2(-7)$$
   $$3+6(-2) = 5+(-14)$$
   $$3+(-12) = -9$$
   $$-9 = -9$$
   When n is replaced with -7, the equation balances, so -7 is the correct solution.

4. Solve this equation and check the solution: $5+8(m+2) = 5m+6$.

   _____

Name: Date:
Instructor: Section:

Objective 2 Practice Exercises

For extra help, see Examples 3–4 on pages 144–145 of your text.

Solve each equation and check.

4. $40 = 5(b + 8)$

4. _____

5. $t + 28 - 10 = 4(t + 6) - 24$

5. _____

6. $-87 + 3g = 9(g - 5) + 8g$

6. _____

Name: Date:
Instructor: Section:

Chapter 3 SOLVING APPLICATION PROBLEMS

3.1 Problem Solving: Perimeter

Learning Objectives
1. Use the formula for perimeter of a square to find the perimeter or the length of one side.
2. Use the formula for perimeter of a rectangle to find the perimeter, the length, or the width.
3. Find the perimeter of parallelograms, triangles, and irregular shapes.

Key Terms

Use the vocabulary terms listed below to complete each statement in exercises 1−6.

formula   perimeter   square

rectangle   parallelogram   triangle

1. A figure with exactly three sides is a _____.

2. The distance around the outside edges of a flat shape is the _____.

3. A _____ is a rule for solving common types of problems.

4. A figure with four sides that are all the same length and meet to form right angles is a _____.

5. A four-sided figure in which all sides meet to form right angles is a _____.

6. A four-sided figure in which opposite sides are both parallel and equal in length is a _____.

Objective 1 Use the formula for perimeter of a square to find the perimeter or the length of one side.

Video Examples

Review these examples for Objective 1:	Now Try:
1. Find the perimeter of the square that measures 7 ft on each side. Use the formula for perimeter of a square, $P = 4s$. Replace s with 7 ft. $P = 4s$ $P = 4 \cdot 7$ ft $P = 28$ ft The perimeter of the square is 28 ft.	1. Find the perimeter of the square that measures 15 in. on each side. _____

Name: Date:
Instructor: Section:

2. If the perimeter of a square is 120 ft, find the length of one side.

Use the formula for perimeter of a square, $P = 4s$. Replace P with 120 ft.
$$P = 4s$$
$$120 \text{ ft} = 4s$$
$$\frac{120 \text{ ft}}{4} = \frac{4s}{4}$$
$$30 \text{ ft} = s$$
The length of one side of the square is 30 ft.

2. If the perimeter of a square is 500 cm, find the length of one side.

_____

Objective 1 Practice Exercises

For extra help, see Examples 1–2 on page 167 of your text.

Find the perimeter of each square, using the appropriate formula.

1.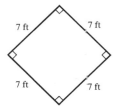

1. _____

2.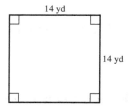

2. _____

For the given perimeter of the square, find the length of one side using the appropriate formula.

3. The perimeter is 76 miles.

3. _____

Name: Date:
Instructor: Section:

Objective 2 Use the formula for perimeter of a rectangle to find the perimeter, the length, or the width.

Video Examples

Review these examples for Objective 2:

3. Find the perimeter of a rectangle with length 32 m and width 17 m.

   The replace l with 32 m and w with 17 m.
   $P = 2l + 2w$
   $P = 2 \cdot 32$ m $+ 2 \cdot 17$ m
   $P = 64$ m $+ 34$ m
   $P = 98$ m
   The perimeter of the rectangle is 98 m.

   Check
   To check the solution, add the lengths of the four sides.
   $P = 32$ m $+ 32$ m $+ 17$ m $+ 17$ m
   $P = 98$ m

4. If the perimeter of a rectangle is 32 ft and the width is 4 ft, find the length.

   Use the formula $P = 2l + 2w$. Replace P with 32 ft and w with 4 ft.
   $P = 2l + 2w$
   32 ft $= 2l + 2 \cdot 4$ ft
   32 ft $= 2l + 8$ ft
   $\underline{-8\text{ ft} \qquad -8\text{ ft}}$
   24 ft $= 2l$
   $\dfrac{24\text{ ft}}{2} = \dfrac{2l}{2}$
   12 ft $= l$
   The length is 12 ft.

Now Try:

3. Find the perimeter of a rectangle with length 54 ft and width 38 ft.

   _____

4. If the perimeter of a rectangle is 48 in. and the width is 6 in., find the length.

   _____

Objective 2 Practice Exercises

For extra help, see Examples 3–4 on pages 168–169 of your text.

Find the perimeter of the rectangle, using the appropriate formula.

4.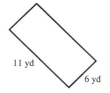
11 yd
6 yd

4. _____

Name: Date:
Instructor: Section:

For each rectangle, you are given the perimeter and either the length or the width. Find the unknown measurement by using the appropriate formula.

5.   The perimeter is 90 ft and the length is 32 ft.   5. _____

6.   The width is 17 cm and the perimeter is 76 cm.   6. _____

Objective 3   Find the perimeter of parallelograms, triangles, and irregular shapes.

Video Examples

Review these examples for Objective 3:   **Now Try:**

5. Find the perimeter of the parallelogram with adjacent sides 15 m and 7 m.

    To find the perimeter, add the lengths of the sides.
    $P = 15\text{ m} + 15\text{ m} + 7\text{ m} + 7\text{ m}$
    $P = 44\text{ m}$

5. Find the perimeter of the parallelogram with adjacent sides 25 m and 12 m.

    _____

7. Find the perimeter of the figure.

    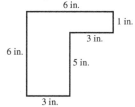

    Find the perimeter by adding the lengths of the sides.
    $P = 6\text{ in.} + 1\text{ in.} + 3\text{ in.} + 5\text{ in.} + 3\text{ in.} + 6\text{ in.}$
    $P = 24\text{ in.}$

7. Find the perimeter of the figure.

    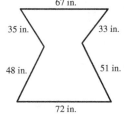

    _____

Name: Date:
Instructor: Section:

Objective 3 Practice Exercises

For extra help, see Examples 5–7 on page 170 of your text.

Find the perimeter of each shape.

7. Parallelogram

7. _____

8.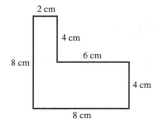

8. _____

9.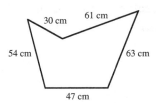

9. _____

Name: Date:
Instructor: Section:

Chapter 3 SOLVING APPLICATION PROBLEMS

3.2 Problem Solving: Area

Learning Objectives
1. Use the formula for area of a rectangle to find the area, the length, or the width.
2. Use the formula for area of a square to find the area or the length of one side.
3. Use the formula for area of a parallelogram to find the area, the base, or the height.
4. Solve application problems involving perimeter and area of rectangles, squares, or parallelograms.

Key Terms

Use the vocabulary terms listed below to complete each statement in exercises 1–4.

   area   rectangle   square   parallelogram

1. To find the area of a _____, use the formula $A = s^2$.

2. To find the area of a _____, use the formula $A = bh$.

3. To find the area of a _____, use the formula $A = lw$.

4. The surface inside a two-dimensional (flat) shape is the _____.

Objective 1   Use the formula for area of a rectangle to find the area, the length, or the width.

Video Examples

Review these examples for Objective 1:
1b. Find the area of the rectangle.

 A rectangle measuring 14 cm by 8 cm.

 The length of this rectangle is 14 cm and the width is 8 cm. Then use the formula $A = lw$.
  $A = l \cdot w$
  $A = 14 \text{ cm} \cdot 8 \text{ cm}$
  $A = 112 \text{ cm}^2$

The area of the rectangle is 112 cm².

Now Try:
1b. Find the area of the rectangle.

 A rectangle measuring 27 cm by 7 cm.

 _____

Name: Date:
Instructor: Section:

2. If the area of a rectangular rug is 24 yd², and the length is 6 yd, find the width.

Use the formula for area of a rectangle, $A = lw$. Replace A with 24 yd², and replace l with 6 yd.

$$A = l \cdot w$$

$$24 \text{ yd}^2 = 6 \text{ yd} \cdot w$$

$$\frac{24 \text{ yd} \cdot \cancel{\text{yd}}}{6 \cancel{\text{yd}}} = \frac{6 \text{ yd} \cdot w}{6 \text{ yd}}$$

$$4 \text{ yd} = w$$

The width of the rug is 4 yd.

Check   To check the solution, use the area formula. Replace l with 6 yd and replace w with 4 yd.

$$A = l \cdot w$$

$$A = 6 \text{ yd} \cdot 4 \text{ yd}$$

$$A = 24 \text{ yd}^2$$

An area of 24 yd² matches the information in the original problem. So 4 yd is the correct width of the rug.

Now Try:
2. If the area of a rectangular flower garden is 35 ft², and the length is 7 ft, find the width.

_____

Objective 1 Practice Exercises

For extra help, see Examples 1–2 on pages 176–177 of your text.

Find the area of each rectangle using the appropriate formula.

1.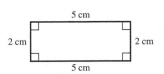

1. _____

2.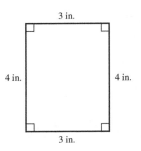

2. _____

Name: Date:
Instructor: Section:

Use the area of the rectangle and either its length or width, and the appropriate formula, to find the other measurement.

3. The area of a food tray is 4104 cm^2 and the width is 54 cm. Find its length.

3. _____

Objective 2 Use the formula for area of a square to find the area or the length of one side.

Video Examples

Review this example for Objective 2:
3. Find the area of a square that is 5 ft on each side.

Use the formula for area of a square, $A = s^2$.
Replace s with 5 ft.

$A = s^2$

$A = s \cdot s$

$A = 5 \text{ ft} \cdot 5 \text{ ft}$

$A = 25 \text{ ft}^2$

The area of the square is 25 ft^2.

Now Try:
3. Find the area of a square that is 13 in. on each side.

_____

Objective 2 Practice Exercises

For extra help, see Examples 3–4 on pages 177–178 of your text.

Find the area of each square using the appropriate formula.

4.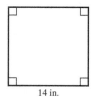
14 in.

4. _____

Name: Date:
Instructor: Section:

5. 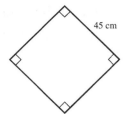   5. _____

Given the area of the square, find the length of one side by inspection.

6.   The area of a square pot holder is 49 in.^{2}.   6. _____

Objective 3   Use the formula for area of a parallelogram to find the area, the base, or the height.

Video Examples

Review these examples for Objective 3:
5a. Find the area of the parallelogram.

The base is 32 cm and the height is 15 cm.

Use the formula for the area of a parallelogram, $A = bh$. Replace b with 32 cm and replace h with 15 cm.

$A = b \cdot h$

$A = 32 \text{ cm} \cdot 15 \text{ cm}$

$A = 480 \text{ cm}^2$

The area of the parallelogram is 480 cm^2.

Now Try:
5a. Find the area of the parallelogram.
The base is 19 cm and the height is 9 cm.

_____

Name: Date:
Instructor: Section:

6. The area of a parallelogram is 48 ft² and the base is 8 ft. Find the height.

Use the formula for the area of a parallelogram, $A = bh$. The value of A is 48 ft², and the value of b is 8 ft.

$$A = b \cdot h$$
$$48 \text{ ft}^2 = 8 \text{ ft} \cdot h$$
$$\frac{48 \text{ ft} \cdot \cancel{\text{ft}}}{8 \cancel{\text{ft}}} = \frac{8 \text{ ft} \cdot h}{8 \text{ ft}}$$
$$6 \text{ ft} = h$$

The height of the parallelogram is 6 ft.

Check To check the solution use the area formula.
$$A = b \cdot h$$
$$A = 8 \text{ ft} \cdot 6 \text{ ft}$$
$$A = 48 \text{ ft}^2$$

An area of 48 ft² matches the information in the original problem. So 6 ft is the correct height of the parallelogram.

6. The area of a parallelogram is 54 ft² and the base is 6 ft. Find the height.

_____

Objective 3 Practice Exercises

For extra help, see Examples 5–6 on pages 179–180 of your text.

Find the area of each parallelogram using the appropriate formula.

7.

7. _____

8.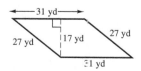

8. _____

Name: Date:
Instructor: Section:

Use the area of the parallelogram and either its base or height, and the appropriate formula, to find the other measurement.

9. The area is 72 m², and the height is 12 m. Find the base.

9. _____

Objective 4 Solve application problems involving perimeter and area of rectangles, squares, or parallelograms.

Video Examples

Review this example for Objective 4:

7. Carla plans to adorn her afghan with a ribbon border. If the afghan measures 7 ft long and 5 ft wide, and the ribbon costs $3 per linear foot, how much is the total cost for this decoration?

The ribbon will go around the edges of the afghan, so you need to find the perimeter of the afghan. Replace l with 7 ft and replace w with 5 ft.

$P = 2l + 2w$

$P = 2 \cdot 7 \text{ ft} + 2 \cdot 5 \text{ ft}$

$P = 14 \text{ ft} + 10 \text{ ft}$

$P = 24 \text{ ft}$

The perimeter of the afghan is 24 ft, so Carla will need 24 ft of ribbon. The cost of the ribbon is $3 per linear foot, which means $3 for 1 foot. To find the cost for 24 ft, multiply $3 \cdot 24$ ft. The ribbon will cost $72.

Now Try:

7. A soybean field is 332 m wide and 512 m long. Find the perimeter and area of the field.

_____

Name: Date:
Instructor: Section:

Objective 4 Practice Exercises

For extra help, see Example 7 on page 180 of your text.

Solve each application problem. You may need to find the perimeter, the area, or one of the side measurements

10. Chuck and Carla plan to carpet their 8 m by 4 m living room. If carpeting costs $9 per square meter, how much will carpeting this room cost?

10. _____

11. Tabitha plans to sew a quilt measuring 88 ft^2 in area. If the quilt must be 11 ft long, how wide will the quilt be?

11. _____

12. A rectangular dance floor measures 27 yd long and 15 yd wide. To cover the floor in wood paneling, it would cost $12 per square yard. How much is the total cost for covering this dance floor in wood paneling?

12. _____

Name: Date:
Instructor: Section:

Chapter 3 SOLVING APPLICATION PROBLEMS

3.3   Solving Application Problems with One Unknown Quantity

Learning Objectives
1   Translate word phrases into algebraic expressions.
2   Translate sentences into equations.
3   Solve application problems with one unknown quantity.

Key Terms

Use the vocabulary terms listed below to complete each statement in exercises 1–4.

sum   **difference**   **product**   **quotient**
increased by   **less than**   **double**   **per**

1. _____ and _____ are words that mean addition.

2. _____ and _____ are words that mean multiplication.

3. _____ and _____ are words that mean division.

4. _____ and _____ are words that mean subtraction.

Objective 1   Translate word phrases into algebraic expressions.

Video Examples

Review these examples for Objective 1:	Now Try:
1. Write each phrase as an algebraic expression. Use x as the variable.	1. Write each phrase as an algebraic expression. Use x as the variable.
a. A number plus 15	**a.** A number plus 7
Algebraic expression: $x + 15$ or $15 + x$	_____
b. The sum of 18 and a number	**b.** The sum of 43 and a number
Algebraic expression: $18 + x$ or $x + 18$	_____
c. 110 more than a number	**c.** 63 more than a number
Algebraic expression: $x + 110$ or $110 + x$	_____
d. −66 added to a number	**d.** −20 added to a number
Algebraic expression: $-66 + x$ or $x + (-66)$	_____

Name: Date:
Instructor: Section:

    e. A number increased by 39

    Algebraic expression: $x + 39$ or $39 + x$

    f. 17 less than a number

    Algebraic expression: $x - 17$

    g. A number subtracted from 73

    Algebraic expression: $73 - x$

    h. 73 subtracted from a number

    Algebraic expression: $x - 73$

    i. 34 fewer than a number

    Algebraic expression: $x - 34$

    j. A number decreased by 51

    Algebraic expression: $x - 51$

    k. 39 minus a number

    Algebraic expression: $39 - x$

2. Write each phrase as an algebraic expression. Use x as the variable.

    a. 112 times a number

    Algebraic expression: $112x$

    b. The product of 25 and a number

    Algebraic expression: $25x$

    c. Triple a number

    Algebraic expression: $3x$

    d. The quotient of -7 and a number

    Algebraic expression: $\dfrac{-7}{x}$

    e. A number divided by 9

    Algebraic expression: $\dfrac{x}{9}$

---

    e. A number increased by 75

    _____

    f. 44 less than a number

    _____

    g. A number subtracted from 89

    _____

    h. 89 subtracted from a number

    _____

    i. 16 fewer than a number

    _____

    j. A number decreased by 36

    _____

    k. 48 minus a number

    _____

2. Write each phrase as an algebraic expression. Use x as the variable.

    a. 57 times a number

    _____

    b. The product of 43 and a number

    _____

    c. Double a number

    _____

    d. The quotient of -13 and a number

    _____

    e. A number divided by 19

    _____

f. 17 subtracted from 8 times a number

Algebraic expression: $8x - 17$

f. 81 subtracted from 6 times a number

_____

Objective 1 Practice Exercises

For extra help, see Examples 1–2 on page 187 of your text.

Write an algebraic expression using x as the variable.

1. The product of –6 and a number

   1. _____

2. The quotient of a number and 10

   2. _____

3. One more than three times a number

   3. _____

Objective 2   Translate sentences into equations.

Video Examples

Review this example for Objective 2:

3. If 6 times a number is added to 13, the result is 37. Find the number. Let *x* represent the number.

   Let *x* represent the unknown number. Use the information in the problem to write an equation.

   <u>6 times a number</u>   <u>added to</u>   13 is 37.
   $$\downarrow \qquad\qquad \downarrow \quad \downarrow\downarrow\downarrow$$
   $$6x \qquad\qquad + \quad 13 = 37$$

   Next, solve the equation.
   $$6x + 13 = 37$$
   $$\underline{-13 \quad -13}$$
   $$6x + 0 = 24$$
   $$\frac{6x}{6} = \frac{24}{6}$$
   $$x = 4$$

   The number is 4.

   Check   Go back to the words of the original problem.

   6 times <u>a number</u> <u>added to</u> 13 is 37.
   $$\downarrow\downarrow \quad\; \downarrow \qquad\quad \downarrow \quad\; \downarrow\downarrow\downarrow$$
   $$6 \;\cdot \quad\; 4 \qquad\quad + \quad 13 = 37$$

   Since $6 \cdot 4 + 13 = 24 + 13 = 37$, then 4 is the correct solution.

Now Try:

3. If 7 times a number is added to 21, the result is 56. Write the equation and find the number.

   _____

Name: Date:
Instructor: Section:

Objective 2 Practice Exercises

For extra help, see Example 3 on page 188 of your text.

Translate each sentence into an equation and solve it. Check your solution by going back to the words in the original problem.

4. If three times a number is decreased by eight, the result is 58. Find the number.

4. Equation _____

Solution _____

5. The sum of three and seven times a number is 31. Find the number.

5. Equation _____

Solution _____

6. If seven times a number is subtracted from twelve times the number, the result is −30. Find the number.

6. Equation _____

Solution _____

Name: Date:
Instructor: Section:

Objective 3   Solve application problems with one unknown quantity.

Video Examples

Review this example for Objective 3:

5. An office supply store bought nine boxes of black ink printer cartridges. One day the store sold 13 cartridges and the next day the store sold 22 cartridges. If nineteen cartridges were left on the shelf, how many cartridges were in each box?

Step 1 Read the problem. Find the number of cartridges in each box.

Step 2 Assign a variable. There is only one unknown quantity. Let n represent the number of cartridges in each box.

Step 3 Write an equation.

Number of boxes	cartridges in each box	took out 13	took out 22	Ended with 19
9	· n	−13	−22	=19

Step 4 Solve.

$$9n - 13 - 22 = 19$$
$$9n + (-13) + (-22) = 19$$
$$9n + (-35) = 19$$
$$+35 +35$$
$$\overline{9n + 0 = 54}$$
$$\frac{9n}{9} = \frac{54}{9}$$
$$n = 6$$

Step 5 State the answer. Each box contains 6 cartridges.

Step 6 Check the solution.
9 boxes each contains 6 cartridges, so $9 \cdot 6 = 54$.
13 were sold, so $54 - 13 = 41$.
Then 22 were sold, so $41 - 22 = 19$.
Ended with 19. This checks.

Now Try:

5. Before a camping weekend, Angela bought 4 packages of paper plates. On Saturday, one package was used up while on Sunday 30 paper plates were used. If two packages plus 25 plates remain, how many paper plates were there in each package?

_____

Name: Date:
Instructor: Section:

Objective 3 Practice Exercises

For extra help, see Examples 4–6 on pages 189–191 of your text.

Solve each application problem. Use the six problem-solving steps listed in your text.

7. The number of bananas an adult gorilla ate is 15 less than three times the number a younger gorilla ate. If the adult gorilla ate 21 bananas, how many did the younger gorilla eat?

7. _____

8. A local jeweler sold three gold watches for the same retail price and established a checking account with this money. The jeweler then bought a diamond for $4000 and a pearl necklace for $11,000 with money from this account. The jeweler now realized this new account was overdrawn for $6000. For how much did she sell each of these gold watches?

8. _____

9. There were 12 celery sticks in the vegetable crisper in Lance's refrigerator. After eating some of these celery sticks, Lance added 16 carrot sticks to the crisper. Later, Lance noticed there were a total of 21 celery and carrot sticks altogether. How many celery sticks did Lance eat?

9. _____

Name: Date:
Instructor: Section:

Chapter 3 SOLVING APPLICATION PROBLEMS

3.4 Solving Application Problems with Two Unknown Quantities

Learning Objectives
1 Solve application problems with two unknown quantities.

Key Terms

Use the vocabulary terms listed below to complete each statement in exercises 1–4.

added to **subtracted from** **times** **divided by**
minus **more than** **half** **triple**

1. _____ and _____ are words that mean addition.

2. _____ and _____ are words that mean multiplication.

3. _____ and _____ are words that mean division.

4. _____ and _____ are words that mean subtraction.

Objective 1 Solve application problems with two unknown quantities.

Video Examples

Review these examples for Objective 1:

3. A site of a new building is rectangular in shape. The length is twice the width. If it will require 642 meters of fence to enclose the site, find the length and the width of the new building site.

   Step 1 Read the problem. Find the length and the width.

   Step 2 Assign a variable. There are two unknowns. Let x represent the width. Let $2x$ represent the length.

   Step 3 Write an equation. Use the formula for perimeter of a rectangle, $P = 2l + 2w$. Replace P with 642, replace l with $2x$, and replace w with x.

   $P = 2\ l\ + 2w$
   $642 = 2(2x) + 2x$

Now Try:

3. The length of a picture frame is 5 in. less than three times the width. If the perimeter of the frame is 86 in., find the length and the width of the picture frame.

   _____

Name: Date:
Instructor: Section:

Step 4 Solve.
$$642 = 2(2x) + 2x$$
$$642 = 4x + 2x$$
$$\frac{642}{6} = \frac{6x}{6}$$
$$107 = x$$

Step 5 State the answer. The width is 107 m. 2x represents the length. Replace x with 107. $2 \cdot 107 = 214$, so the length is 214 m.

Step 6 Check the solution.
$$P = 2 \cdot 214 + 2 \cdot 107$$
$$P = 428 + 214$$
$$P = 642$$
This matches the perimeter given in the original problem.

2. A string is 89 cm long. Bob's cat, Reginald, bit the string into two pieces so that one piece is 17 cm longer than the other. Find the length of each piece.

Step 1 Read the problem. Find the length of a longer piece and a shorter piece.

Step 2 Assign a variable. There are two unknowns. Let x represent the length of the shorter piece. Let $x + 17$ represent the length of the longer piece.

Step 3 Write an equation.

shorter piece		longer piece		Total length
x	+	x+17	=	89

Step 4 Solve.
$$x + x + 17 = 89$$
$$2x + 17 = 89$$
$$-17-17$$
$$2x + 0 = 72$$
$$\frac{2x}{2} = \frac{72}{2}$$
$$x = 36$$

2. A telephone cable 98 meters in length is cut into two pieces. If one piece is 22 meters longer than the other, how long are the two pieces?

_____

Name: Date:
Instructor: Section:

Step 5 State the answer. The shorter piece is 36 cm.
$x + 17$ represents the longer piece. Replace x with 36.
$36 + 17 = 53$, so the longer piece is 53 cm.

Step 6 Check the solution. The total length is 36 cm + 53 cm = 89 cm. This checks.

Objective 1 Practice Exercises

For extra help, see Examples 1–3 on pages 196–199 of your text.

Solve each application problem by using the six problem-solving steps.

1. The vote totals for the two candidates for county judge, Jones and Greer, differed by only 516 votes, with Jones receiving more votes. If there were 16,786 voters in the county, how many votes did each candidate receive?

    1. Greer_____

       Jones_____

2. After a 14-meter tall tree is chopped down, it is cut into four pieces. Three pieces are the same length, while the fourth piece is 2 m longer than each of the other three. Find the length of each piece.

    2. _____

Name: Date:
Instructor: Section:

Use the formula $P = 2l + 2w$ to solve. Make a sketch to help you solve the problem.

3. A rectangular garden is three times as long as it is wide. The perimeter of the garden is 96 yd. Find the length and the width of the garden.

3. width_____

length _____

Name: Date:
Instructor: Section:

Chapter 4 RATIONAL NUMBERS: POSITIVE AND NEGATIVE FRACTIONS

4.1 Introduction to Signed Fractions

Learning Objectives	
1	Use a fraction to name part of a whole.
2	Identify numerators, denominators, proper fractions, and improper fractions.
3	Graph positive and negative fractions on a number line.
4	Find the absolute value of a fraction.
5	Write equivalent fractions.

Key Terms

Use the vocabulary terms listed below to complete each statement in exercises 1−6.

fraction   **numerator**   **denominator**

proper fraction   **improper fraction**   **equivalent fractions**

1. Two fractions are _____ when they represent the same portion of a whole.

2. A number of the form $\frac{a}{b}$, where *a* and *b* are integers and $b \neq 0$ is called a _____.

3. A fraction whose numerator is larger than its denominator is called an _____.

4. In the fraction $\frac{2}{9}$, the 2 is the _____.

5. A fraction whose denominator is larger than its numerator is called a _____.

6. The _____ of a fraction shows the number of equal parts in a whole.

Name: Date:
Instructor: Section:

Objective 1 Use a fraction to name part of a whole.

Video Examples

Review these examples for Objective 1:

1. Use fractions to represent the shaded portion and the unshaded portion of the figure.

   The figure has 8 equal parts. The 3 shaded parts are represented by the fraction $\frac{3}{8}$. The unshaded part is $\frac{5}{8}$ of the figure.

2. Use a fraction to represent the shaded parts.

   An area equal to 5 of the $\frac{1}{2}$ parts is shaded, so $\frac{5}{2}$ is shaded.

Now Try:

1. Use fractions to represent the shaded portion and the unshaded portion of the figure.

   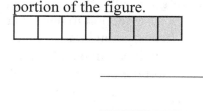

   _____

   _____

2. Use a fraction to represent the shaded parts.

   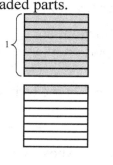

   _____

Objective 1 Practice Exercises

For extra help, see Examples 1–2 on pages 218–219 of your text.

Write the fractions that represent the shaded and unshaded portions of each figure.

1.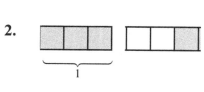

   1. Shaded _____

      Unshaded _____

2.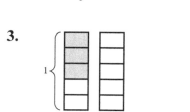

   2. Shaded _____

      Unshaded _____

3.

   3. Shaded _____

      Unshaded _____

Name: Date:
Instructor: Section:

Objective 2 Identify numerators, denominators, proper fractions, and improper fractions.

Video Examples

Review these examples for Objective 2:

3b. Identify the numerator and denominator in the fraction. Then state the number of equal parts in the whole.

$$\frac{4}{15}$$

$\frac{4}{15}$ ← Numerator
← Denominator

15 equal parts in the whole

4. $\frac{4}{7}, \frac{7}{10}, \frac{19}{6}, \frac{11}{11}, \frac{13}{27}, \frac{1}{8}, \frac{6}{5}$

a. Identify all the proper fractions in the list.

Proper fractions have a numerator that is less than the denominator.

$\frac{4}{7}, \frac{7}{10}, \frac{13}{27}, \frac{1}{8}$

b. Identify all the improper fractions in the list.

Improper fractions have a numerator that is equal or greater than the denominator.

$\frac{19}{6}, \frac{11}{11}, \frac{6}{5}$

Now Try:

3b. Identify the numerator and denominator in the fraction. Then state the number of equal parts in the whole.

$$\frac{6}{23}$$

_____

_____

_____

4. $\frac{7}{8}, \frac{9}{11}, \frac{22}{7}, \frac{13}{13}, \frac{12}{29}, \frac{1}{5}, \frac{17}{4}$

a. Identify all the proper fractions in the list.

_____

b. Identify all the improper fractions in the list.

_____

Objective 2 Practice Exercises

For extra help, see Examples 3–4 on pages 219–220 of your text.

Identify the numerator and denominator in each fraction.

4. $\frac{2}{9}$

4. numerator _____

denominator _____

5. $\frac{12}{5}$

5. numerator _____

denominator _____

Copyright © 2018 Pearson Education, Inc.

Name: Date:
Instructor: Section:

List the proper and improper fractions in the group of numbers.

6. $\frac{2}{3}, \frac{5}{2}, \frac{4}{4}, \frac{6}{7}, \frac{8}{15}, \frac{20}{19}$

6. Proper _____

   Improper _____

Objective 3 Graph positive and negative fractions on a number line.

Video Examples

Review this example for Objective 3:

5. Graph the fractions on the number line.

   $-\frac{7}{9}, \frac{7}{9}$

   For $-\frac{7}{9}$, the fraction is negative, so it is between 0 and −1. We divide that space into 9 equal parts. Then we start at 0 and count to the left 7 parts.

   There is no sign in front of $\frac{7}{9}$, so it is positive.

   Because $\frac{7}{9}$, is between 0 and 1, we divide that space into 9 equal parts. Then we start at 0 and count to the right 7 parts.

   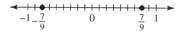

Now Try:

5. Graph the fractions on the number line.

   $\frac{5}{6}, -\frac{5}{6}$

Objective 3 Practice Exercises

For extra help, see Example 5 on page 221 of your text.

Graph each pair of fractions on a number line.

7. $\frac{2}{3}, -\frac{2}{3}$

7.
![number line from -1 to 1]

8. $-\frac{1}{4}, \frac{1}{4}$

8.
![number line from -1 to 1]

9. $-\frac{5}{8}, \frac{5}{8}$

9.
![number line from -1 to 1]

84

Name: Date:
Instructor: Section:

Objective 4 Find the absolute value of a fraction.

Video Examples

Review this example for Objective 4:

6. Find each absolute value: $\left|-\frac{3}{5}\right|$ and $\left|\frac{3}{5}\right|$.

The distance from 0 to $-\frac{3}{5}$ on the number line is $\frac{3}{5}$ space, so $\left|-\frac{3}{5}\right| = \frac{3}{5}$.

The distance from 0 to $\frac{3}{5}$ on the number line is also $\frac{3}{5}$ space, so $\left|\frac{3}{5}\right| = \frac{3}{5}$.

Now Try:

6. Find each absolute value: $\left|\frac{9}{10}\right|$ and $\left|-\frac{9}{10}\right|$.

_____

_____

Objective 4 Practice Exercises

For extra help, see Example 6 on page 221 of your text.

Find each absolute value.

10. $\left|-\frac{15}{7}\right|$

11. $\left|\frac{8}{9}\right|$

12. $|0|$

10. _____

11. _____

12. _____

Objective 5 Write equivalent fractions.

Video Examples

Review these examples for Objective 5:

7.
  a. Write $-\frac{1}{3}$ as an equivalent fraction with a denominator of 18.

  In other words, $-\frac{1}{3} = -\frac{?}{18}$.

  The original denominator is 3. Multiplying 3 times 6 gives 18, the new denominator. To write

Now Try:

7.
  a. Write $-\frac{4}{5}$ as an equivalent fraction with a denominator of 30.

  _____

Name: Date:
Instructor: Section:

an equivalent fraction, multiply both the numerator and denominator by 6.
$$-\frac{1}{3} = -\frac{1 \cdot 6}{3 \cdot 6} = -\frac{6}{18}$$
So, $-\frac{1}{3}$ is equivalent to $-\frac{6}{18}$.

b. Write $\frac{15}{18}$ as an equivalent fraction with a denominator of 6.

In other words, $\frac{15}{18} = \frac{?}{6}$.

The original denominator is 18. Dividing 18 by 3 gives 6, the new denominator. To write an equivalent fraction, divide both the numerator and denominator by 3.
$$\frac{15}{18} = \frac{15 \div 3}{18 \div 3} = \frac{5}{6}$$
So, $\frac{15}{18}$ is equivalent to $\frac{5}{6}$.

b. Write $\frac{35}{42}$ as an equivalent fraction with a denominator of 6.

_____

Objective 5 Practice Exercises

For extra help, see Examples 7–8 on pages 223–224 of your text.

Rewrite each fraction as an equivalent fraction with a denominator of 48.

13. $-\frac{2}{3}$ 13. _____

14. $\frac{3}{4}$ 14. _____

Rewrite the fraction as an equivalent fraction with a denominator of 5.

15. $-\frac{4}{20}$ 15. _____

Name: Date:
Instructor: Section:

Chapter 4 RATIONAL NUMBERS: POSITIVE AND NEGATIVE FRACTIONS

4.2 Writing Fractions in Lowest Terms

Learning Objectives
1   Identify fractions written in lowest terms.
2   Write a fraction in lowest terms using common factors.
3   Write a number as a product of prime factors.
4   Write a fraction in lowest terms using prime factorization.
5   Write a fraction with variables in lowest terms.

Key Terms

Use the vocabulary terms listed below to complete each statement in exercises 1–4.

lowest terms   prime number   composite number

prime factorization

1. A _____ has at least one factor other than itself and 1.

2. In a _____ every factor is a prime number.

3. The factors of a _____ are itself and 1.

4. A fraction is written in _____ when its numerator and denominator have no common factor other than 1.

Objective 1   Identify fractions written in lowest terms.

Video Examples

Review these examples for Objective 1:

1. Are the following fractions in lowest terms? If not, find a common factor of the numerator and denominator (other than 1).

   a. $\frac{3}{7}$

   The numerator and denominator have no common factor other than 1, so the fraction is in lowest terms.

   b. $\frac{28}{49}$

   The numerator and denominator have a common factor of 7, so the fraction is not in lowest terms.

Now Try:

1. Are the following fractions in lowest terms? If not, find a common factor of the numerator and denominator (other than 1).

   a. $\frac{8}{9}$

   _____

   b. $\frac{27}{36}$

   _____

Name: Date:
Instructor: Section:

Objective 1 Practice Exercises

For extra help, see Example 1 on page 229 of your text.

Are the following fractions in lowest terms? If not, find a common factor of the numerator and denominator (other than 1).

1. $-\dfrac{7}{11}$ 1. _____

2. $\dfrac{20}{25}$ 2. _____

3. $\dfrac{12}{16}$ 3. _____

Objective 2   Write a fraction in lowest terms using common factors.

Video Examples

Review these examples for Objective 2:

2. Divide by a common factor to write each fraction in lowest terms.

   a. $\dfrac{40}{48}$

   The greatest common factor of 40 and 48 is 8. Divide both numerator and denominator by 8.
   $$\dfrac{40}{48} = \dfrac{40 \div 8}{48 \div 8} = \dfrac{5}{6}$$

   b. $\dfrac{35}{55}$

   Divide both numerator and denominator by 5.
   $$\dfrac{35}{55} = \dfrac{35 \div 5}{55 \div 5} = \dfrac{7}{11}$$

   c. $-\dfrac{28}{49}$

   Divide both numerator and denominator by 7. Keep the negative sign.
   $$-\dfrac{28}{49} = -\dfrac{28 \div 7}{49 \div 7} = -\dfrac{4}{7}$$

Now Try:

2. Divide by a common factor to write each fraction in lowest terms.

   a. $\dfrac{20}{36}$

   _____

   b. $\dfrac{35}{40}$

   _____

   c. $-\dfrac{25}{65}$

   _____

Name: Date:
Instructor: Section:

d. $\frac{80}{96}$ **d.** $\frac{55}{60}$

Dividing by 4 does not give the greatest common factor of 80 and 96.

$\frac{80}{96} = \frac{80 \div 4}{96 \div 4} = \frac{20}{24}$ ← not lowest terms

There is another common factor of 4.

$\frac{20}{24} = \frac{20 \div 4}{24 \div 4} = \frac{5}{6}$ ← lowest terms

Divide by 16, the greatest common factor, in one step.

$\frac{80}{96} = \frac{80 \div 16}{96 \div 16} = \frac{5}{6}$

Objective 2 Practice Exercises

For extra help, see Example 2 on page 230 of your text.

Write in lowest terms.

4. $-\frac{24}{36}$ 4. _____

5. $\frac{30}{42}$ 5. _____

6. $-\frac{32}{72}$ 6. _____

Objective 3   Write a number as a product of prime factors.

Video Examples

Review these examples for Objective 3:

3. Label each number as prime or composite or neither.
    1 3 4 6 17 23

    First, 1 is neither prime nor composite.
    Next, 3, 17, and 23 are prime.
    The numbers 4 and 6 can be divided by 2, so 4 and 6 are composite.

Now Try:

3. Label each number as prime or composite or neither.
    1 8 21 29 31

    _____

Name: Date:
Instructor: Section:

4a. Find the prime factorization of the number.

54

Start by dividing 54 by 2 and work your way up the chain of divisions.

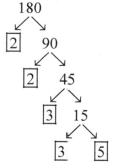

Because all the factors (divisors) are prime, the prime factorization of 54 is $54 = 2 \cdot 3 \cdot 3 \cdot 3$.

5. Find the prime factorization of each number.

a. 180

Try to divide 180 by 2. The quotient is 90. Write the factors 2 and 90 under 180. Box the 2 (or circle it), because it is prime.

```
   180
   ↙ ↘
  [2]  90
```

Try dividing 90 by 2. The quotient is 45. Write the factors 2 and 45 under the 90. Continue until you find the prime factorization.

```
   180
   ↙ ↘
  [2]  90
       ↙ ↘
      [2]  45
           ↙ ↘
          [3]  15
               ↙ ↘
              [3] [5]
```

The prime factorization is $180 = 2 \cdot 2 \cdot 3 \cdot 3 \cdot 5$.

4a. Find the prime factorization of the number.
96

_____

5. Find the prime factorization of each number.
a. 152

_____

b. 108

Divide 108 by 2, the first prime number.

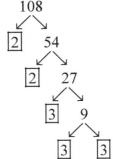

The prime factorization is $108 = 2 \cdot 2 \cdot 3 \cdot 3 \cdot 3$.

b. 168

_____

Objective 3 Practice Exercises

For extra help, see Examples 3–5 on pages 231–233 of your text.

Find the prime factorization of each number.

7. 105

7. _____

8. 126

8. _____

9. 320

9. _____

Name: Date:
Instructor: Section:

Objective 4 Write a fraction in lowest terms using prime factorization.

Video Examples

Review these examples for Objective 4:
6. Write each fraction in lowest terms.

   a. $\dfrac{28}{63}$

   Write the prime factorizations of 28 and 63.
   $$\dfrac{28}{63} = \dfrac{2 \cdot 2 \cdot 7}{3 \cdot 3 \cdot 7}$$
   Divide the numerator and denominator by 7, the common factor.
   $$\dfrac{28}{63} = \dfrac{2 \cdot 2 \cdot \cancel{7}}{3 \cdot 3 \cdot \cancel{7}} = \dfrac{4}{9}$$

   b. $\dfrac{90}{126}$

   Write the prime factorizations of 90 and 126. Then divide the numerator and denominator by the common factors.
   $$\dfrac{90}{126} = \dfrac{\cancel{2} \cdot \cancel{3} \cdot \cancel{3} \cdot 5}{\cancel{2} \cdot \cancel{3} \cdot \cancel{3} \cdot 7} = \dfrac{5}{7}$$

Now Try:
6. Write each fraction in lowest terms.

   a. $\dfrac{75}{225}$

   _____

   b. $\dfrac{42}{140}$

   _____

Objective 4 Practice Exercises

For extra help, see Example 6 on pages 234–235 of your text.

Write each numerator and denominator as a product of prime factors. Then use the prime factorization to write the fraction in lowest terms.

10. $\dfrac{18}{99}$ 10. _____

11. $\dfrac{63}{105}$ 11. _____

Name: Date:
Instructor: Section:

12. $\dfrac{36}{210}$ 12. _____

Objective 5 Write a fraction with variables in lowest terms.

Video Examples

Review these examples for Objective 5:
7. Write each fraction in lowest terms.

  a. $\dfrac{10}{15x}$

  $\dfrac{10}{15x} = \dfrac{2 \cdot \overset{1}{\cancel{5}}}{3 \cdot \underset{1}{\cancel{5}} \cdot x} = \dfrac{2}{3x}$

  b. $\dfrac{7xy}{28xy}$

  $\dfrac{7xy}{28xy} = \dfrac{\overset{1}{\cancel{7}} \cdot \overset{1}{\cancel{x}} \cdot \overset{1}{\cancel{y}}}{2 \cdot 2 \cdot \underset{1}{\cancel{7}} \cdot \underset{1}{\cancel{x}} \cdot \underset{1}{\cancel{y}}} = \dfrac{1}{4}$

  c. $\dfrac{6b^4}{12ab^2}$

  $\dfrac{6b^4}{12ab^2} = \dfrac{\overset{1}{\cancel{2}} \cdot \overset{1}{\cancel{3}} \cdot b \cdot b \cdot \overset{1}{\cancel{b}} \cdot \overset{1}{\cancel{b}}}{\underset{1}{\cancel{2}} \cdot 2 \cdot \underset{1}{\cancel{3}} \cdot a \cdot \underset{1}{\cancel{b}} \cdot \underset{1}{\cancel{b}}} = \dfrac{b^2}{2a}$

Now Try:
7. Write each fraction in lowest terms.

  a. $\dfrac{25}{30x}$

  _____

  b. $\dfrac{8ab}{24ab}$

  _____

  c. $\dfrac{9b^5}{36ab^3}$

  _____

Name: Date:
Instructor: Section:

Objective 5 Practice Exercises

For extra help, see Example 7 on page 236 of your text.

Write each fraction in lowest terms.

13. $\dfrac{12r^2s}{4rs^3}$ 13. _____

14. $\dfrac{16b^2cd}{40b^2d}$ 14. _____

15. $\dfrac{8xy^2}{6x^2y^2}$ 15. _____

Name: Date:
Instructor: Section:

Chapter 4 RATIONAL NUMBERS: POSITIVE AND NEGATIVE FRACTIONS

4.3 Multiplying and Dividing Signed Fractions

Learning Objectives
1. Multiply signed fractions.
2. Multiply fractions that involve variables.
3. Divide signed fractions.
4. Divide fractions that involve variables.
5. Solve application problems involving multiplying and dividing fractions.

Key Terms

Use the vocabulary terms listed below to complete each statement in exercises 1−4.

    reciprocals   indicator words   of   each

1. Two numbers are _____ of each other if their product is 1.

2. The words "per" and "divided equally" are _____ for division.

3. When the word "_____" follows a fraction, it means "multiply".

4. The word "_____" indicates division.

Objective 1 Multiply signed fractions.

Video Examples

Review these examples for Objective 1:	**Now Try:**
1. Find each product.	1. Find each product.
a. $-\dfrac{7}{9} \cdot -\dfrac{5}{11}$	a. $-\dfrac{10}{11} \cdot -\dfrac{4}{13}$
Multiply the numerators and multiply the denominators.
$-\dfrac{7}{9} \cdot -\dfrac{5}{11} = \dfrac{7 \cdot 5}{9 \cdot 11} = \dfrac{35}{99}$	_____
The answer is in lowest terms because 35 and 99 have no common factor other than 1.
b. $\left(\dfrac{3}{5}\right)\left(-\dfrac{4}{7}\right)$	b. $\left(\dfrac{15}{22}\right)\left(-\dfrac{3}{4}\right)$
$\left(\dfrac{3}{5}\right)\left(-\dfrac{4}{7}\right) = -\dfrac{3 \cdot 4}{5 \cdot 7} = -\dfrac{12}{35}$	_____

Copyright © 2018 Pearson Education, Inc.

Name: Date:
Instructor: Section:

2b. Multiply. Write the product in lowest terms.

Find $\dfrac{3}{8}$ of $\dfrac{4}{9}$.

Recall that "of" indicates multiplication.

$$\dfrac{3}{8} \cdot \dfrac{4}{9} = \dfrac{3 \cdot 2 \cdot 2}{2 \cdot 2 \cdot 2 \cdot 3 \cdot 3} = \dfrac{\cancel{3} \cdot \cancel{2} \cdot \cancel{2}}{\cancel{2} \cdot \cancel{2} \cdot 2 \cdot \cancel{3} \cdot 3} = \dfrac{1}{6}$$

2b. Multiply. Write the product in lowest terms.

Find $\dfrac{2}{7}$ of $\dfrac{21}{40}$.

_____

Objective 1 Practice Exercises

For extra help, see Examples 1–3 on pages 241–244 of your text.

Multiply. Write the products in lowest terms.

1. $-\dfrac{10}{42} \cdot \dfrac{3}{5}$

1. _____

2. $\dfrac{6}{18} \cdot \dfrac{9}{2}$

2. _____

3. $\dfrac{5}{9}$ of 81

3. _____

Objective 2   Multiply fractions that involve variables.

Video Examples

Review these examples for Objective 2:
4. Find each product.

  a. $\dfrac{6x}{7} \cdot \dfrac{5}{18x}$

$$\dfrac{6x}{7} \cdot \dfrac{5}{18x} = \dfrac{2 \cdot 3 \cdot x \cdot 5}{7 \cdot 2 \cdot 3 \cdot 3 \cdot x} = \dfrac{\cancel{2} \cdot \cancel{3} \cdot \cancel{x} \cdot 5}{7 \cdot \cancel{2} \cdot \cancel{3} \cdot 3 \cdot \cancel{x}} = \dfrac{5}{21}$$

Now Try:
4. Find each product.

  a. $\dfrac{12x}{5} \cdot \dfrac{7}{18x}$

_____

Name: Date:
Instructor: Section:

b. $\left(\dfrac{9a}{5b}\right)\left(\dfrac{10b^2}{3a}\right)$ **b.** $\left(\dfrac{7x}{9c}\right)\left(\dfrac{6c^2}{35x}\right)$

$\left(\dfrac{9a}{5b}\right)\left(\dfrac{10b^2}{3a}\right) = \dfrac{3\cdot 3\cdot a\cdot 2\cdot 5\cdot b\cdot b}{5\cdot b\cdot 3\cdot a}$

$\phantom{\left(\dfrac{9a}{5b}\right)\left(\dfrac{10b^2}{3a}\right)} = \dfrac{\cancel{3}\cdot 3\cdot \cancel{a}\cdot 2\cdot \cancel{5}\cdot \cancel{b}\cdot b}{\cancel{5}\cdot \cancel{b}\cdot \cancel{3}\cdot \cancel{a}}$

$\phantom{\left(\dfrac{9a}{5b}\right)\left(\dfrac{10b^2}{3a}\right)} = 6b$

Objective 2 Practice Exercises

For extra help, see Example 4 on page 244 of your text.

Use prime factorization to find these products.

4. $\dfrac{3c}{5} \cdot \dfrac{c}{9}$ 4. _____

5. $\left(\dfrac{m}{10}\right)\left(\dfrac{25}{m^2}\right)$ 5. _____

6. $\dfrac{5a}{7b^2} \cdot \dfrac{14b}{10a^2}$ 6. _____

Name: Date:
Instructor: Section:

Objective 3 Divide signed fractions.

Video Examples

Review this example for Objective 3:

5a. Rewrite the division problem as a multiplication problem.

$$\frac{5}{7} \div \frac{10}{3}$$

$$\frac{5}{7} \div \frac{10}{3} = \frac{5}{7} \cdot \frac{3}{10} = \frac{\overset{1}{\cancel{5}} \cdot 3}{7 \cdot 2 \cdot \underset{1}{\cancel{5}}} = \frac{3}{14}$$

Now Try:

5a. Rewrite the division problem as a multiplication problem.

$$\frac{5}{9} \div \frac{20}{3}$$

_____

Objective 3 Practice Exercises

For extra help, see Example 5 on pages 246–247 of your text.

Divide. Write the quotients in lowest terms.

7. $\frac{7}{8} \div (-21)$

7. _____

8. $-\frac{5}{12} \div \frac{15}{8}$

8. _____

9. $-\frac{2}{3} \div \left(-\frac{7}{9}\right)$

9. _____

Name: Date:
Instructor: Section:

Objective 4   Divide fractions that involve variables.

Video Examples

Review these examples for Objective 4:   **Now Try:**

6. Divide. (Assume that none of the variables represent zero.)   6. Divide. (Assume that none of the variables represent zero.)

     b. $\dfrac{9b^2}{7} \div b^3$ b. $\dfrac{10b}{9} \div b^3$

$$\dfrac{9b^2}{7} \div b^3 = \dfrac{9b^2}{7} \cdot \dfrac{1}{b^3} = \dfrac{3 \cdot 3 \cdot \cancel{b} \cdot \cancel{b} \cdot 1}{7 \cdot \cancel{b} \cdot \cancel{b} \cdot b} = \dfrac{9}{7b}$$

     a. $\dfrac{x^3}{y^2} \div \dfrac{x^2}{4y}$ a. $\dfrac{a^2}{b^3} \div \dfrac{a}{5b^2}$

$$\dfrac{x^3}{y^2} \div \dfrac{x^2}{4y} = \dfrac{x^3}{y^2} \cdot \dfrac{4y}{x^2} = \dfrac{\cancel{x} \cdot \cancel{x} \cdot x \cdot 2 \cdot 2 \cdot \cancel{y}}{\cancel{y} \cdot y \cdot \cancel{x} \cdot \cancel{x}} = \dfrac{4x}{y}$$

Objective 4 Practice Exercises

For extra help, see Example 6 on page 247 of your text.

Divide. Write the quotients in lowest terms.

10. $\dfrac{3b}{5a} \div \dfrac{6}{7ab}$ 10. _____

11. $10x^2 \div \dfrac{5x}{3}$ 11. _____

12. $\dfrac{7c}{3d} \div 14c^2 d$ 12. _____

Name: Date:
Instructor: Section:

Objective 5   Solve application problems involving multiplying and dividing fractions.

Video Examples

Review this example for Objective 5:

7a. Solve the application problem.

Signe gives $\frac{1}{12}$ of her income to charities. Last year she earned $48,000. How much did she give to charities?

Because the word "of" follows the fraction, it indicates multiplication.

$$\frac{1}{12} \cdot 48,000 = \frac{1}{12} \cdot \frac{48,000}{1} = \frac{1 \cdot \cancel{12} \cdot 4000}{\cancel{12} \cdot 1} = 4000$$

She gave $4000 to charities.

Now Try:

7a. Solve the application problem.

A flower bed is $\frac{12}{11}$ m by $\frac{7}{8}$ m. Find its area.

_____

Objective 5 Practice Exercises

For extra help, see Example 7 on page 248 of your text.

Solve each application problem.

13. Phillip saves $\frac{2}{9}$ of his income each month. How much did he save last month if he earned $2700?

13. _____

14. Rochelle sells hats at craft shows. She needs $\frac{2}{5}$ yd for each hat. How many hats can she make from 10 yards of fabric?

14. _____

15. How many $\frac{1}{3}$-lb servings will a 24-lb turkey provide?

15. _____

Name: Date:
Instructor: Section:

Chapter 4 RATIONAL NUMBERS: POSITIVE AND NEGATIVE FRACTIONS

4.4   Adding and Subtracting Signed Fractions

Learning Objectives
1   Add and subtract like fractions.
2   Find the lowest common denominator for unlike fractions.
3   Add and subtract unlike fractions.
4   Add and subtract unlike fractions that contain variables.

Key Terms

Use the vocabulary terms listed below to complete each statement in exercises 1–3.

   like fractions   **unlike fractions**   **least common denominator**

1. Fractions with different denominators are called _____.

2. Fractions with the same denominator are called _____.

3. The _____ of two whole numbers is the smallest whole number divisible by both of the numbers.

Objective 1   Add and subtract like fractions.

Video Examples

Review these examples for Objective 1:   **Now Try:**
1. Find each sum or difference.   1. Find each sum or difference.

   a. $\dfrac{1}{9} + \dfrac{5}{9}$ a. $\dfrac{1}{12} + \dfrac{7}{12}$

   These are like fractions because they have a common denominator. Add the numerators and write the sum over the common denominator.

   $\dfrac{1}{9} + \dfrac{5}{9} = \dfrac{1+5}{9} = \dfrac{6}{9}$

   Now write $\dfrac{6}{9}$ in lowest terms.

   $\dfrac{6}{9} = \dfrac{2 \cdot \cancel{3}^{1}}{3 \cdot \cancel{3}_{1}} = \dfrac{2}{3}$

   b. $-\dfrac{4}{7} + \dfrac{6}{7}$ b. $-\dfrac{1}{4} + \dfrac{3}{4}$

   $-\dfrac{4}{7} + \dfrac{6}{7} = \dfrac{-4+6}{7} = \dfrac{2}{7}$

Copyright © 2018 Pearson Education, Inc.   101

Name: Date:
Instructor: Section:

c. $\dfrac{4}{15} - \dfrac{7}{15}$

Write the subtraction as adding the opposite.

$\dfrac{4}{15} - \dfrac{7}{15} = \dfrac{4-7}{15} = \dfrac{4+(-7)}{15} = \dfrac{-3}{15}$, or $-\dfrac{3}{15}$

Now write $-\dfrac{3}{15}$ in lowest terms.

$-\dfrac{3}{15} = -\dfrac{\cancel{3}}{\cancel{3} \cdot 5} = -\dfrac{1}{5}$

d. $\dfrac{6}{x^3} - \dfrac{4}{x^3}$

$\dfrac{6}{x^3} - \dfrac{4}{x^3} = \dfrac{6-4}{x^3} = \dfrac{2}{x^3}$

c. $\dfrac{7}{20} - \dfrac{19}{20}$

_____

d. $\dfrac{9}{a^2} - \dfrac{5}{a^2}$

_____

Objective 1 Practice Exercises

For extra help, see Example 1 on page 254 of your text.

Write each sum or difference in lowest terms.

1. $-\dfrac{14}{15} + \dfrac{4}{15}$

1. _____

2. $\dfrac{3}{4a} + \dfrac{1}{4a}$

2. _____

3. $\dfrac{7}{y^2} - \dfrac{3}{y^2}$

3. _____

Name: Date:
Instructor: Section:

Objective 2   Find the lowest common denominator for unlike fractions.

Video Examples

Review these examples for Objective 2:

2.

    a. Find the LCD for $\frac{2}{7}$ and $\frac{5}{21}$ by inspection.

Since 21 is divisible by 7, then 21 is the LCD.

    b. Find the LCD for $\frac{7}{12}$ and $\frac{3}{8}$ by inspection.

Since 12 is not divisible by 8, use multiples of 12, that is 12, 24, and 36. Notice that 24 is divisible by both 8 and 12. The LCD is 24.

3.

    a. What is the LCD for $\frac{11}{18}$ and $\frac{5}{24}$?

Write 18 and 24 as the product of prime factors.
$18 = 2 \cdot 3 \cdot 3$
$24 = 2 \cdot 2 \cdot 2 \cdot 3$
$LCD = 2 \cdot 2 \cdot 2 \cdot 3 \cdot 3 = 72$
The LCD for $\frac{11}{18}$ and $\frac{5}{24}$ is 72.

    b. What is the LCD for $\frac{9}{16}$ and $\frac{7}{36}$?

$16 = 2 \cdot 2 \cdot 2 \cdot 2$
$36 = 2 \cdot 2 \cdot 3 \cdot 3$
$LCD = 2 \cdot 2 \cdot 2 \cdot 2 \cdot 3 \cdot 3 = 144$
The LCD for $\frac{9}{16}$ and $\frac{7}{36}$ is 144.

Now Try:

2.

    a. Find the LCD for $\frac{2}{9}$ and $\frac{7}{27}$ by inspection.

_____

    b. Find the LCD for $\frac{3}{10}$ and $\frac{7}{15}$ by inspection.

_____

3.

    a. What is the LCD for $\frac{9}{20}$ and $\frac{8}{25}$?

_____

    b. What is the LCD for $\frac{7}{30}$ and $\frac{2}{45}$?

_____

Objective 2 Practice Exercises

For extra help, see Examples 2–3 on page 256 of your text.

Find the LCD for each pair of fractions.

    4. $\frac{3}{7}$ and $\frac{3}{14}$ **4.** _____

Name: Date:
Instructor: Section:

5. $\frac{8}{21}$ and $\frac{8}{9}$

5. _____

6. $\frac{7}{12}$ and $\frac{3}{40}$

6. _____

Objective 3 Add and subtract unlike fractions.

Video Examples

Review these examples for Objective 3:
4. Find each sum or difference.

   c. $-\frac{7}{18} + \frac{5}{12}$

   Step 1 Use prime factorization to find the LCD.
   $18 = 2 \cdot 3 \cdot 3$
   $12 = 2 \cdot 2 \cdot 3$
   $LCD = 2 \cdot 2 \cdot 3 \cdot 3 = 36$
   Step 2
   $-\frac{7}{18} = -\frac{7 \cdot 2}{18 \cdot 2} = -\frac{14}{36}$ and $\frac{5}{12} = \frac{5 \cdot 3}{12 \cdot 3} = \frac{15}{36}$
   Step 3 Add the numerators.
   $-\frac{7}{18} + \frac{5}{12} = -\frac{14}{36} + \frac{15}{36} = \frac{-14+15}{36} = \frac{1}{36}$
   Step 4 $\frac{1}{36}$ is in lowest terms.

   d. $5 - \frac{3}{4}$

   Step 1 The LCD for $\frac{5}{1}$ and $\frac{3}{4}$ is 4, the larger denominator.
   Step 2 $\frac{5}{1} = \frac{5 \cdot 4}{1 \cdot 4} = \frac{20}{4}$ and $\frac{3}{4}$ already has the LCD.
   Step 3 Subtract the numerators.
   $\frac{5}{1} - \frac{3}{4} = \frac{20}{4} - \frac{3}{4} = \frac{20-3}{4} = \frac{17}{4}$
   Step 4 $\frac{17}{4}$ is in lowest terms.

Now Try:
4. Find each sum or difference.

   c. $-\frac{5}{24} + \frac{7}{9}$

   _____

   d. $9 - \frac{2}{7}$

   _____

Name: Date:
Instructor: Section:

b. $\dfrac{3}{8} - \dfrac{7}{12}$

Step 1 The LCD is 24.

Step 2 $\dfrac{3}{8} = \dfrac{3 \cdot 3}{8 \cdot 3} = \dfrac{9}{24}$ and $\dfrac{7}{12} = \dfrac{7 \cdot 2}{12 \cdot 2} = \dfrac{14}{24}$

Step 3 Subtract the numerators.
$\dfrac{3}{8} - \dfrac{7}{12} = \dfrac{9}{24} - \dfrac{14}{24} = \dfrac{9-14}{24} = \dfrac{-5}{24}$, or $-\dfrac{5}{24}$

Step 4 $-\dfrac{5}{24}$ is in lowest terms.

b. $\dfrac{8}{15} - \dfrac{7}{10}$

_____

Objective 3 Practice Exercises

For extra help, see Example 4 on pages 257–258 of your text.

Find each sum or difference. Write all answers in lowest terms.

7. $\dfrac{1}{6} + \dfrac{2}{15}$

7. _____

8. $-\dfrac{1}{2} + \dfrac{7}{12}$

8. _____

9. $\dfrac{33}{40} - \dfrac{7}{24}$

9. _____

Name: Date:
Instructor: Section:

Objective 4   Add and subtract unlike fractions that contain variables.

Video Examples

Review these examples for Objective 4:
5. Find each sum or difference.

   a. $\dfrac{1}{3}+\dfrac{x}{2}$

   Step 1   The LCD is 6.
   Step 2 $\dfrac{1}{3}=\dfrac{1\cdot 2}{3\cdot 2}=\dfrac{2}{6}$ and $\dfrac{x}{2}=\dfrac{x\cdot 3}{2\cdot 3}=\dfrac{3x}{6}$
   Step 3 $\dfrac{1}{3}+\dfrac{x}{2}=\dfrac{2}{6}+\dfrac{3x}{6}=\dfrac{2+3x}{6}$
   Step 4 $\dfrac{2+3x}{6}$ is in lowest terms.

   b. $\dfrac{5}{6}-\dfrac{7}{x}$

   Step 1   The LCD is $6x$.
   Step 2 $\dfrac{5}{6}=\dfrac{5\cdot x}{6\cdot x}=\dfrac{5x}{6x}$ and $\dfrac{7}{x}=\dfrac{7\cdot 6}{x\cdot 6}=\dfrac{42}{6x}$
   Step 3 $\dfrac{5}{6}-\dfrac{7}{x}=\dfrac{5x}{6x}-\dfrac{42}{6x}=\dfrac{5x-42}{6x}$
   Step 4 $\dfrac{5x-42}{6x}$ is in lowest terms.

Now Try:
5. Find each sum or difference.

   a. $\dfrac{2}{3}+\dfrac{x}{5}$

   _____

   b. $\dfrac{5}{8}-\dfrac{8}{x}$

   _____

Objective 4 Practice Exercises

For extra help, see Example 5 on pages 258–259 of your text.

Find each sum or difference. Write all answers in lowest terms.

10. $\dfrac{3}{n}+\dfrac{3}{5}$ 10. _____

11. $\dfrac{1}{6}+\dfrac{x}{3}$ 11. _____

12. $\dfrac{3}{a}-\dfrac{b}{4}$ 12. _____

Name: Date:
Instructor: Section:

Chapter 4 RATIONAL NUMBERS: POSITIVE AND NEGATIVE FRACTIONS

4.5   Problem Solving: Mixed Numbers and Estimating

Learning Objectives
1   Identify mixed numbers and graph them on a number line.
2   Rewrite mixed numbers as improper fractions, or the reverse.
3   Estimate the answer and multiply or divide mixed numbers.
4   Estimate the answer and add or subtract mixed numbers.
5   Solve application problems containing mixed numbers.

Key Terms

Use the vocabulary terms listed below to complete each statement in exercises 1–2.

    mixed number   **improper fraction**

1. A(n) _____ includes a fraction and a whole number written together.

2. A mixed number can be rewritten as a(n) _____.

Objective 1   Identify mixed numbers and graph them on a number line.

Video Examples

Review this example for Objective 1:

1. Use a number line to show mixed numbers $\frac{5}{2}$ and $-\frac{5}{2}$.

    $\frac{5}{2}$ is equivalent to $2\frac{1}{2}$.

    $-\frac{5}{2}$ is equivalent to $-2\frac{1}{2}$.

    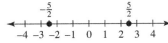

Now Try:

1. Use a number line to show mixed numbers $1\frac{1}{4}$ and $-1\frac{1}{4}$.

    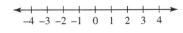

Name: Date:
Instructor: Section:

Objective 1 Practice Exercises

For extra help, see Example 1 on page 266 of your text.

Graph the mixed numbers or improper fractions on a number line.

1. $-3\frac{1}{2}$ and $3\frac{1}{2}$

    1.
    ```
    <-+--+--+--+--+--+--+--+--+->
      -4 -3 -2 -1  0  1  2  3  4
    ```

2. $\frac{10}{3}$ and $-\frac{10}{3}$

    2.
    ```
    <-+--+--+--+--+--+--+--+--+->
      -4 -3 -2 -1  0  1  2  3  4
    ```

3. $-\frac{11}{4}$ and $\frac{11}{4}$

    3.
    ```
    <-+--+--+--+--+--+--+--+--+->
      -4 -3 -2 -1  0  1  2  3  4
    ```

Objective 2 Rewrite mixed numbers as improper fractions, or the reverse.

Video Examples

Review these examples for Objective 2:

2. Write $9\frac{3}{4}$ as an improper fraction.

    Step 1 $\quad 9\frac{3}{4} \quad 4 \cdot 9 = 36 \quad$ Then $36 + 3 = 39$

    Step 2 $\quad 9\frac{3}{4} = \frac{39}{4}$

3b. Write the improper fraction as an equivalent mixed number in simplest form.

    $\frac{22}{6}$

    Divide 22 by 6.

    $$6\overline{)22} \quad \text{so } \frac{22}{6} = 3\frac{4}{6} = 3\frac{2}{3}$$
    $$\underline{18}$$
    $$4$$

    Or, write $\frac{22}{6}$ in lowest terms first.

    $\frac{22}{6} = \frac{\cancel{2} \cdot 11}{\cancel{2} \cdot 3} = \frac{11}{3} \quad$ Then $3\overline{)11} \quad$ so $\frac{11}{3} = 3\frac{2}{3}$
    $\underline{9}$
    2

Now Try:

2. Write $5\frac{5}{6}$ as an improper fraction.

    _____

3b. Write the improper fraction as an equivalent mixed number in simplest form.

    $\frac{37}{9}$

    _____

Name: Date:
Instructor: Section:

Objective 2 Practice Exercises

For extra help, see Examples 2–3 on pages 267–269 of your text.

Write each mixed number as an improper fraction.

4. $-8\frac{2}{7}$ 4. _____

5. $-1\frac{7}{9}$ 5. _____

Write the improper fraction as a mixed number in simplest form.

6. $\frac{26}{3}$ 6. _____

Objective 3 Estimate the answer and multiply or divide mixed numbers.

Video Examples

Review these examples for Objective 3:

5a. First, round the numbers and estimate the answer. Then find the exact answer. Write the exact answer in simplest form.

$$2\frac{1}{3} \cdot 4\frac{5}{7}$$

Estimate the answer by rounding the mixed numbers.

$2\frac{1}{3}$ rounds to 2   and $4\frac{5}{7}$ rounds to 5

$2 \cdot 5 = 10 \leftarrow$ Estimated answer

To find the exact answer, first rewrite each mixed number as an improper fraction.

$2\frac{1}{3} = \frac{7}{3}$ and $4\frac{5}{7} = \frac{33}{7}$

Next, multiply.

$$2\frac{1}{3} \cdot 4\frac{5}{7} = \frac{7}{3} \cdot \frac{33}{7} = \frac{\cancel{7} \cdot \cancel{3} \cdot 11}{\cancel{3} \cdot \cancel{7}} = \frac{11}{1} = 11$$

The estimate was 10, so an exact answer of 11 is reasonable.

Now Try:

5a. First, round the numbers and estimate the answer. Then find the exact answer. Write exact the answer in simplest form.

$$4\frac{9}{10} \cdot 3\frac{4}{7}$$

Estimate _____

Exact _____

Name: Date:
Instructor: Section:

6a. First, round the numbers and estimate the answer. Then find the exact answer. Write the exact answer in simplest form.

$$3\frac{1}{5} \div 2\frac{2}{3}$$

Estimate the answer by rounding the mixed numbers.

$3\frac{1}{5}$ rounds to 3   and $2\frac{2}{3}$ rounds to 3

$3 \div 3 = 1 \leftarrow$ Estimated answer

To find the exact answer, first rewrite each mixed number as an improper fraction.

$$3\frac{1}{5} \div 2\frac{2}{3} = \frac{16}{5} \div \frac{8}{3}$$

Now rewrite the problem as multiplying by the reciprocal of $\frac{8}{3}$.

$$\frac{16}{5} \div \frac{8}{3} = \frac{16}{5} \cdot \frac{3}{8} = \frac{\cancel{8} \cdot 2 \cdot 3}{5 \cdot \cancel{8}} = \frac{6}{5} = 1\frac{1}{5}$$

The estimate was 1, so an exact answer of $1\frac{1}{5}$ is reasonable.

6a. First, round the numbers and estimate the answer. Then find the exact answer. Write the exact answer in simplest form.

$$4\frac{4}{9} \div 6\frac{2}{3}$$

Estimate _____

Exact _____

Objective 3 Practice Exercises

For extra help, see Examples 4–6 on page 269–271 of your text.

First, round the mixed numbers to the nearest whole number and estimate each answer; then find the exact answer. Write exact answers in simplest form. Write each mixed number as an improper fraction.

7. $3\frac{2}{3} \cdot 1\frac{1}{7}$

7.
Estimate_____

Exact _____

Name: Date:
Instructor: Section:

8. $3\frac{3}{4} \div 1\frac{3}{8}$

8.
Estimate_____

Exact _____

9. $5\frac{1}{3} \div 2\frac{2}{5}$

9.
Estimate_____

Exact _____

Objective 4 Estimate the answer and add or subtract mixed numbers.

Video Examples

Review these examples for Objective 4:

7. First, estimate each answer. Then add or subtract to find the exact answer. Write exact answers in simplest form.

   a. $3\frac{3}{10} + 4\frac{4}{5}$

   To estimate the answer, round each mixed number to the nearest whole number.

   $3\frac{3}{10} + 4\frac{4}{5}$

   $3 \;+\; 5 = 8 \;\leftarrow$ Estimate

   To find the exact answer, first rewrite each mixed number as an equivalent improper fraction.

   $3\frac{3}{10} + 4\frac{4}{5} = \frac{33}{10} + \frac{24}{5}$

   The LCD for $\frac{33}{10}$ and $\frac{24}{5}$ is 10. Rewrite $\frac{24}{5}$ as an equivalent fraction with a denominator of 10.

   $\frac{33}{10} + \frac{24}{5} = \frac{33}{10} + \frac{48}{10} = \frac{33+48}{10} = \frac{81}{10} = 8\frac{1}{10}$

Now Try:

7. First, estimate each answer. Then add or subtract to find the exact answer. Write exact answers in simplest form.

   a. $1\frac{2}{9} + 5\frac{1}{3}$

   Estimate _____

   Exact _____

Copyright © 2018 Pearson Education, Inc.

Name: Date:
Instructor: Section:

The estimate was 8, so an exact answer of $8\frac{1}{10}$ is reasonable.

b. $5\frac{1}{4} - 3\frac{5}{6}$

Round each number and estimate the answer.
$5\frac{1}{4} - 3\frac{5}{6}$
$5 - 4 = 1 \leftarrow$ Estimate

To find the exact answer, rewrite the mixed numbers as improper fractions and subtract.

$5\frac{1}{4} - 3\frac{5}{6} = \frac{21}{4} - \frac{23}{6} = \frac{63}{12} - \frac{46}{12}$
$= \frac{63-46}{12} = \frac{17}{12} = 1\frac{5}{12}$

The estimate was 1, so an exact answer of $1\frac{5}{12}$ is reasonable.

c. $7 - 3\frac{2}{5}$

$7 - 3\frac{2}{5}$
$7 - 3 = 4 \leftarrow$ Estimate
$7 - 3\frac{2}{5} = \frac{7}{1} - \frac{17}{5} = \frac{35}{5} - \frac{17}{5}$
$= \frac{35-17}{5} = \frac{18}{5} = 3\frac{3}{5}$

The estimate was 4, so an exact answer of $3\frac{3}{5}$ is reasonable.

b. $12\frac{5}{12} - 8\frac{5}{6}$

Estimate _____

Exact _____

c. $9 - 4\frac{1}{7}$

Estimate _____

Exact _____

Objective 4 Practice Exercises

For extra help, see Example 7 on pages 272–273 of your text.

First, round the mixed numbers to the nearest whole number and estimate each answer; then find the exact answer. Write exact answers in simplest form. Write each mixed number as an improper fraction.

10. $3\frac{5}{6} + 2\frac{3}{4}$

10.
Estimate_____

Exact _____

Name: Date:
Instructor: Section:

11. $9 - 5\frac{4}{9}$

11.
Estimate _____

Exact _____

12. $9\frac{7}{18} - 3\frac{5}{12}$

12.
Estimate _____

Exact _____

Objective 5 Solve application problems containing mixed numbers.

Video Examples

Review this example for Objective 5:

8a. First, estimate the answer to the application problem. Then find the exact answer.

Thomas worked $25\frac{1}{2}$ hours over the last four days. If he worked the same amount each day, how long was he at work each day?

First, round each mixed number to the nearest whole number.

$25\frac{1}{2}$ rounds to 26

Using the rounded numbers, we divide.

$26 \div 4 = 6\frac{1}{2}$ ← Estimate

To find the exact answer, use the original mixed numbers and divide.

$25\frac{1}{2} \div 4 = \frac{51}{2} \div \frac{4}{1} = \frac{51}{2} \cdot \frac{1}{4} = \frac{51 \cdot 1}{2 \cdot 4} = \frac{51}{8} = 6\frac{3}{8}$

Thomas worked $6\frac{3}{8}$ hr each day. This is close to the estimate of $6\frac{1}{2}$ hr.

Now Try:

8a. First, estimate the answer to the application problem. Then find the exact answer.

Katlyn used $2\frac{3}{8}$ packages of nuts in her salad recipe. Each package has $3\frac{1}{6}$ ounces of nuts. How many ounces of nuts did she use in the recipe?

Estimate _____

Exact _____

Name: Date:
Instructor: Section:

Objective 5 Practice Exercises

For extra help, see Example 8 on page 274 of your text.

First, estimate the answer to each application problem. Then find the exact answer.

13. A living room has dimensions $3\frac{3}{4}$ meters by $3\frac{1}{3}$ meters. What is the area of the room?

   13.
   Estimate_____

   Exact _____

14. Suppose that a pair of pants requires $3\frac{1}{8}$ yd of material. How much material would be needed for 6 pairs of pants?

   14.
   Estimate_____

   Exact _____

15. George's daughter grew $1\frac{1}{3}$ inches last year and $2\frac{1}{5}$ inches this year. How much has her height increased over the two years?

   15.
   Estimate_____

   Exact _____

Name: Date:
Instructor: Section:

Chapter 4 RATIONAL NUMBERS: POSITIVE AND NEGATIVE FRACTIONS

4.6 Exponents, Order of Operations, and Complex Fractions

Learning Objectives
1. Simplify fractions with exponents.
2. Use the order of operations with fractions.
3. Simplify complex fractions.

Key Terms

Use the vocabulary terms listed below to complete each statement in exercises 1–3.

complex fraction   exponent   order of operations

1. For problems or expressions with more than one operation, the _____ tells what to do first, second, and so on, to obtain the correct answer.

2. An _____ tells how many times a number is used as a factor in repeated multiplication.

3. A _____ is a fraction in which the numerator and/or denominator contain one or more fractions.

Objective 1   Simplify fractions with exponents.

Video Examples

Review these examples for Objective 1:
1. Simplify.

   a. $\left(-\frac{1}{5}\right)^3$

   $\left(-\frac{1}{5}\right)^3 = \left(-\frac{1}{5}\right)\left(-\frac{1}{5}\right)\left(-\frac{1}{5}\right) = \frac{1}{25}\left(-\frac{1}{5}\right) = -\frac{1}{125}$

   b. $\left(\frac{4}{5}\right)^3 \left(\frac{5}{8}\right)^2$

   $\left(\frac{4}{5}\right)^3 \left(\frac{5}{8}\right)^2 = \left(\frac{4}{5}\right)\left(\frac{4}{5}\right)\left(\frac{4}{5}\right)\left(\frac{5}{8}\right)\left(\frac{5}{8}\right)$

   $= \frac{\cancel{2}\cdot\cancel{2}\cdot\cancel{4}\cdot\cancel{4}\cdot\cancel{5}\cdot\cancel{5}}{\cancel{5}\cdot\cancel{5}\cdot 5\cdot\cancel{2}\cdot\cancel{4}\cdot\cancel{2}\cdot\cancel{4}}$

   $= \frac{1}{5}$

Now Try:
1. Simplify.

   a. $\left(-\frac{1}{4}\right)^3$

   _____

   b. $\left(\frac{5}{7}\right)^3 \left(\frac{7}{10}\right)^2$

   _____

Name: Date:
Instructor: Section:

Objective 1 Practice Exercises

For extra help, see Example 1 on page 281 of your text.

Simplify.

1. $3\left(-\dfrac{1}{3}\right)^3$

   1. _____

2. $\left(-\dfrac{2}{3}\right)^3\left(-\dfrac{3}{2}\right)^2$

   2. _____

3. $\left(-\dfrac{2}{3}\right)^2\left(\dfrac{1}{2}\right)^2$

   3. _____

Objective 2 Use the order of operations with fractions.

Video Examples

Review these examples for Objective 2:

2. Simplify.

   a. $-\dfrac{1}{2}+\dfrac{2}{3}\left(\dfrac{4}{5}\right)$

   $-\dfrac{1}{2}+\dfrac{2}{3}\left(\dfrac{4}{5}\right) = -\dfrac{1}{2}+\dfrac{2\cdot 4}{3\cdot 5} = -\dfrac{1}{2}+\dfrac{8}{15}$

   $= -\dfrac{15}{30}+\dfrac{16}{30} = \dfrac{-15+16}{30}$

   $= \dfrac{1}{30}$

Now Try:

2. Simplify.

   a. $-\dfrac{1}{6}+\dfrac{2}{3}\left(\dfrac{4}{5}\right)$

   _____

Name: Date:
Instructor: Section:

b. $-4+\left(\dfrac{5}{6}-\dfrac{2}{3}\right)^2$

$-4+\left(\dfrac{5}{6}-\dfrac{2}{3}\right)^2 = -4+\left(\dfrac{5}{6}-\dfrac{4}{6}\right)^2$

$= -4+\left(\dfrac{5-4}{6}\right)^2 = -4+\left(\dfrac{1}{6}\right)^2$

$= -4+\left(\dfrac{1}{36}\right)$

$= \dfrac{-144+1}{36} = -\dfrac{143}{36}$

b. $-1+\left(\dfrac{4}{9}-\dfrac{2}{3}\right)^2$

_____

Objective 2 Practice Exercises

For extra help, see Example 2 on page 282 of your text.

Simplify.

4. $\dfrac{15}{16}-\dfrac{1}{8}-\left(\dfrac{3}{4}\right)^2$

4. _____

5. $\left(\dfrac{3}{2}\right)^2 - \dfrac{3}{10} \div \dfrac{1}{5}$

5. _____

6. $\left(\dfrac{1}{3}\right)^2 - \left(-\dfrac{1}{2}\right)^3$

6. _____

Name: Date:
Instructor: Section:

Objective 3 Simplify complex fractions.

Video Examples

Review this example for Objective 3:

3a. Simplify.

$$\frac{-\frac{6}{7}}{-\frac{5}{14}}$$

Rewrite the complex fraction using the $\div$ symbol for division. Then follow the order of operations.

$$\frac{-\frac{6}{7}}{-\frac{5}{14}} = -\frac{6}{7} \div -\frac{5}{14} = -\frac{6}{7} \cdot -\frac{14}{5}$$

$$= \frac{6 \cdot 2 \cdot \cancel{7}}{\cancel{7} \cdot 5} = \frac{12}{5}, \text{ or } 2\frac{2}{5}$$

Now Try:

3a. Simplify.

$$\frac{-\frac{5}{9}}{-\frac{7}{12}}$$

_____

Objective 3 Practice Exercises

For extra help, see Example 3 on page 283 of your text.

Simplify.

7. $\dfrac{\frac{16}{35}}{-\frac{4}{70}}$ 7. _____

8. $\dfrac{-20}{\frac{2}{5}}$ 8. _____

9. $\dfrac{\left(\frac{2}{3}\right)^2}{\left(-\frac{4}{5}\right)^2}$ 9. _____

Name: Date:
Instructor: Section:

Chapter 4 RATIONAL NUMBERS: POSITIVE AND NEGATIVE FRACTIONS

4.7 Problem Solving: Equations Containing Fractions

Learning Objectives	
1	Use the multiplication property of equality to solve equations containing fractions.
2	Use both the addition and the multiplication properties of equality to solve equations containing fractions.
3	Solve application problems using equations containing fractions.

Key Terms

Use the vocabulary terms listed below to complete each statement in exercises 1−2.

multiplication property of equality

division property of equality

1. The _____ states that both sides of an equation may be divided by the same nonzero number and it will still be balanced.

2. The _____ states that both sides of an equation may be multiplied by the same nonzero number and it will still be balanced.

Objective 1 Use the multiplication property of equality to solve equations containing fractions.

Video Examples

Review these examples for Objective 1:
1. Solve each equation and check each solution.

   a. $\frac{1}{5}b = 12$

   Multiply both sides by $\frac{5}{1}$.

   $$\frac{1}{5}b = 12$$
   $$\frac{5}{1}\left(\frac{1}{5}b\right) = \frac{5}{1}(12)$$
   $$\left(\frac{5}{1} \cdot \frac{1}{5}\right)b = \frac{5}{1}\left(\frac{12}{1}\right)$$
   $$1b = \frac{60}{1}$$
   $$b = 60$$

Now Try:
1. Solve each equation and check each solution.

   a. $\frac{1}{7}b = 3$

   _____

Name: Date:
Instructor: Section:

Check $\frac{1}{5}b = 12$

$\frac{1}{5}(60) = 12$

$\dfrac{1 \cdot \cancel{5} \cdot 12}{\cancel{5}} = 12$

$12 = 12$

The solution is 60.

b. $15 = -\frac{3}{5}x$

Multiply both sides by $-\frac{5}{3}$.

$-\frac{5}{3}(15) = -\frac{5}{3}\left(-\frac{3}{5}x\right)$

$-25 = x$

Check $15 = -\frac{3}{5}x$

$15 = -\frac{3}{5}(-25)$

$15 = \dfrac{3 \cdot \cancel{5} \cdot 5}{\cancel{5}}$

$15 = 15$

The solution is -25.

c. $-\frac{5}{6}n = -\frac{1}{4}$

$-\frac{6}{5}\left(-\frac{5}{6}n\right) = -\frac{6}{5}\left(-\frac{1}{4}\right)$

$n = \dfrac{2 \cdot 3}{5 \cdot 2 \cdot 2}$

$n = \dfrac{3}{10}$

Check $-\frac{5}{6}n = -\frac{1}{4}$

$-\frac{5}{6}\left(\frac{3}{10}\right) = -\frac{1}{4}$

$-\dfrac{\cancel{5} \cdot \cancel{3}}{2 \cdot \cancel{3} \cdot 2 \cdot \cancel{5}} = -\frac{1}{4}$

$-\frac{1}{4} = -\frac{1}{4}$

The solution is $\frac{3}{10}$.

b. $25 = -\frac{5}{6}x$

c. $-\frac{7}{9}n = -\frac{14}{45}$

Name: Date:
Instructor: Section:

Objective 1 Practice Exercises

For extra help, see Example 1 on pages 288–290 of your text.

Solve each equation and check each solution.

1. $-30 = \dfrac{5}{6}b$ 1. _____

2. $\dfrac{2}{9}n = 18$ 2. _____

3. $-\dfrac{8}{9}h = -\dfrac{1}{6}$ 3. _____

Name: Date:
Instructor: Section:

Objective 2   Use both the addition and the multiplication properties of equality to solve equations containing fractions.

Video Examples

Review these examples for Objective 2:

2. Solve each equation and check each solution.

   a. $\frac{1}{5}c + 2 = 5$

   Add the opposite of 2, which is –2, to both sides.

   $$\frac{1}{5}c + 2 = 5$$
   $$\underline{-2 -2}$$
   $$\frac{1}{5}c + 0 = 3$$
   $$\frac{1}{5}c = 3$$
   $$\frac{5}{1}\left(\frac{1}{5}c\right) = \frac{5}{1}\left(\frac{3}{1}\right)$$
   $$c = 15$$

   Check $\frac{1}{5}c + 2 = 5$
   $$\frac{1}{5}(15) + 2 = 5$$
   $$3 + 2 = 5$$
   $$5 = 5$$

   The solution is 15.

   b. $-5 = \frac{7}{8}y + 9$

   To get the variable term by itself, add –9 to both sides.

   $$-5 = \frac{7}{8}y + 9$$
   $$\underline{-9 -9}$$
   $$-14 = \frac{7}{8}y + 0$$
   $$\frac{8}{7}(-14) = \frac{8}{7}\left(\frac{7}{8}y\right)$$
   $$-16 = y$$

Now Try:

2. Solve each equation and check each solution.

   a. $\frac{1}{6}c + 3 = 8$

   ——————

   b. $-9 = \frac{5}{7}y + 6$

   ——————

Name: Date:
Instructor: Section:

Check $-5 = \frac{7}{8}y + 9$

$-5 = \frac{7}{8}(-16) + 9$

$-5 = -14 + 9$

$-5 = -5$

The solution is -16.

Objective 2 Practice Exercises

For extra help, see Example 2 on pages 290–291 of your text.

Solve each equation and check each solution.

4. $\frac{1}{5}s - 15 = -10$
4. _____

5. $-8 = \frac{5}{2}r + 2$
5. _____

6. $-\frac{1}{6}n + 8 = -4$
6. _____

Name: Date:
Instructor: Section:

Objective 3   Solve application problems using equations containing fractions.

Video Examples

Review this example for Objective 3:

3. An expression for the recommended weight of an adult is $\frac{11}{2}$(height in inches) − 220.

   A man weighs 187 pounds. What is his height in inches?

   Step 1 Read the problem. It is about weight and height.

   Step 2 Assign a variable. Let h represent the height.

   Step 3 Write an equation.

   $\frac{11}{2}$(height in inches) − 220 is weight

   $$\frac{11}{2}h - 220 = 187$$

   Step 4 Solve.

   $$\frac{11}{2}h - 220 = 187$$
   $$\phantom{\frac{11}{2}h}\; +220 \quad +220$$
   $$\frac{11}{2}h + 0 = 407$$
   $$\frac{2}{11}\left(\frac{11}{2}h\right) = \frac{2}{11}(407)$$
   $$h = 74$$

   Step 5 State the answer. The man is 74 inches, or 6 ft 2 in.

   Step 6 Check.
   If the height is 74 inches, then find the weight.
   $$\frac{11}{2}(74) - 220 = 407 - 220 = 187$$
   The answer checks.

Now Try:

3. An expression for the recommended weight of an adult is $\frac{11}{2}$(height in inches) − 220.

   A man weighs 220 pounds. What is his height in inches?

   _____

Name: Date:
Instructor: Section:

Objective 3 Practice Exercises

For extra help, see Example 3 on page 292 of your text.

In Exercises 7–9, find each person's age using the six problem-solving steps and this expression for approximate systolic blood pressure: $100 + \frac{age}{2}$. *Assume that all the people have normal blood pressure.*

7. A man has a systolic blood pressure of 110. How old is he?

    7. _____

8. A woman has a systolic blood pressure of 126. How old is she?

    8. _____

9. A man has a systolic blood pressure of 115. How old is he?

    9. _____

Name: Date:
Instructor: Section:

Chapter 4 RATIONAL NUMBERS: POSITIVE AND NEGATIVE FRACTIONS

4.8 Geometry Applications: Area and Volume

Learning Objectives
1   Find the area of triangle.
2   Find the volume of a rectangular solid.
3   Find the volume of a pyramid.

Key Terms

Use the vocabulary terms listed below to complete each statement in exercises 1−2.

         volume   **area**

1. _____ is measured in square units.

2. _____ is measured in cubic units.

Objective 1   **Find the area of triangle.**

Video Examples

Review this example for Objective 1:
1a. Find the area of the triangle.

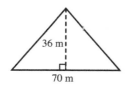

The base is 70 m and the height is 36 m.

$A = \frac{1}{2} \cdot b \cdot h$

$A = \frac{1}{\cancel{2}} \cdot \cancel{70}^{35} \text{ m} \cdot 36 \text{ m}$

$A = 1260 \text{ m}^2$

Now Try:
1a. Find the area of the triangle.

_____

Name: Date:
Instructor: Section:

Objective 1 Practice Exercises

For extra help, see Examples 1–2 on pages 298–299 of your text.

Find the perimeter and area of each triangle.

1.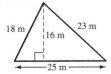

    1. Perimeter_____

       Area _____

2.

    2. Perimeter_____

       Area _____

Find the shaded area in the figure.

3.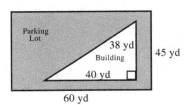

    3. _____

Copyright © 2018 Pearson Education, Inc.

Name: Date:
Instructor: Section:

Objective 2 Find the volume of a rectangular solid.

Video Examples

Review this example for Objective 2:
3b. Find the volume of the box.

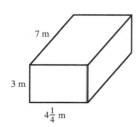

Use the formula $V = l \cdot w \cdot h$.

$V = 7 \text{ m} \cdot 4\frac{1}{4} \text{ m} \cdot 3 \text{ m}$

$V = \frac{7 \text{ m}}{1} \cdot \frac{17 \text{ m}}{4} \cdot \frac{3 \text{ m}}{1}$

$V = \frac{7 \text{ m} \cdot 17 \text{ m} \cdot 3 \text{ m}}{4}$

$V = \frac{357}{4} \text{ m}^3$, or $89\frac{1}{4} \text{ m}^3$

Now Try:
3b. Find the volume of the box.

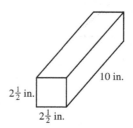

_____

Objective 2 Practice Exercises

For extra help, see Example 3 on page 300 of your text.

Find the volume of each figure.

4.
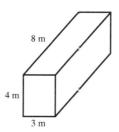

4. _____

5.
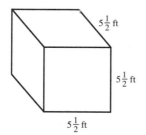

5. _____

Name: Date:
Instructor: Section:

6.

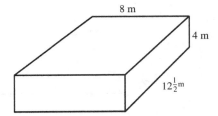

6. _____

Objective 3 Find the volume of a pyramid.

Video Examples

Review this example for Objective 3:

4. Find the volume of this pyramid with rectangular base. Round your answer to the nearest tenth.

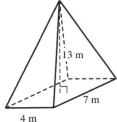

First find the value of B in the formula, which is the area of a rectangular base. Recall that the area of the rectangle is found by multiplying length times width.

$B = 7 \text{ m} \cdot 4 \text{ m}$

$B = 28 \text{ m}^2$

Now find the volume.

$V = \dfrac{B \cdot h}{3}$

$V \approx \dfrac{28 \text{ m}^2 \cdot 13 \text{ m}}{3}$

$V \approx 121.3 \text{ m}^3$

Now Try:

4. Find the volume of this pyramid with rectangular base. Round your answer to the nearest tenth.

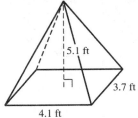

_____

Name: Date:
Instructor: Section:

Objective 3 Practice Exercises

For extra help, see Example 4 on page 301 of your text.

Find the volume of each figure.

7.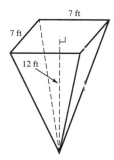

7. _____

8.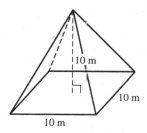

8. _____

9.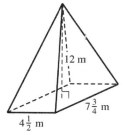

9. _____

Name: Date:
Instructor: Section:

Chapter 5 RATIONAL NUMBERS: POSITIVE AND NEGATIVE DECIMALS

5.1 Reading and Writing Decimal Numbers

Learning Objectives
1   Write parts of a whole using decimals.
2   Identify the place value of a digit.
3   Read decimal numbers.
4   Write decimals as fractions or mixed numbers.

Key Terms

Use the vocabulary terms listed below to complete each statement in exercises 1–3.

    decimals   decimal point   place value

1. We use _____ to show parts of a whole.

2. A _____ is assigned to each place to the left or right of the decimal point.

3. The dot that separates the whole number part from the fractional part of a decimal number is called the _____.

Objective 1   **Write parts of a whole using decimals.**

Objective 1 Practice Exercises

For extra help, see Example 1 on page 325 of your text.

Write the portion of each square that is shaded as a fraction, as a decimal, and in words.

1. 1. _____

                                               _____

                                               _____

2. 2. _____

                                               _____

                                               _____

Name: Date:
Instructor: Section:

3.

3. _____

_____

_____

Objective 2 Identify the place value of a digit.

Video Examples

Review these examples for Objective 2:
2. Identify the place value of each digit.

   a. 486.92

   4 hundreds
   8 tens
   6 ones
   .
   9 tenths
   2 hundredths

   b. 0.00465

   0 ones
   .
   0 tenths
   0 hundredths
   4 thousandths
   6 ten-thousandths
   5 hundred-thousandths

Now Try:
2. Identify the place value of each digit.
   a. 862.93

   _____

   b. 0.00769

   _____

Objective 2 Practice Exercises

For extra help, see Example 2 on page 326 of your text.

Identify the digit that has the given place value.

4. 43.507   tenths   4. _____

              hundredths   _____

5. 2.83714   thousandths   5. _____

              ten-thousandths   _____

132 Copyright © 2018 Pearson Education, Inc.

Name: Date:
Instructor: Section:

Identify the place value of each digit in these decimals.

6. 37.082   3   6. _____

            7   _____

            0   _____

            8   _____

            2   _____

Objective 3   Read decimal numbers.

Video Examples

Review these examples for Objective 3:
3c. Tell how to read the decimal in words.

0.09

Read it as: nine hundredths

4b. Read the decimal.

543.71

five hundred forty-three and seventy-one hundredths

Now Try:
3c. Tell how to read the decimal in words.

0.07

_____

4b. Read the decimal.

18.009

_____

Objective 3 Practice Exercises

For extra help, see Examples 3–4 on pages 326–327 of your text.

Tell how to read each decimal in words.

7. 0.08   7. _____

8. 10.835   8. _____

9. 97.008   9. _____

Name: Date:
Instructor: Section:

Objective 4 Write decimals as fractions or mixed numbers.

Video Examples

Review these examples for Objective 4:

6. Write each decimal as a fraction or mixed number in lowest terms.

   c. 17.216

   $17.216 = 17\dfrac{216}{1000} = 17\dfrac{216 \div 8}{1000 \div 8} = 17\dfrac{27}{125}$

   a. 0.6

   $0.6 = \dfrac{6}{10}$ Write $\dfrac{6}{10}$ in lowest terms.

   $\dfrac{6}{10} = \dfrac{6 \div 2}{10 \div 2} = \dfrac{3}{5}$

Now Try:

6. Write each decimal as a fraction or mixed number in lowest terms.

   c. 6.04

   _____

   a. 0.2

   _____

Objective 4 Practice Exercises

For extra help, see Examples 5–6 on page 328 of your text.

Write each decimal as a fraction or mixed number in lowest terms.

10. 0.001   10. _____

11. 3.6   11. _____

12. 0.95   12. _____

Name: Date:
Instructor: Section:

Chapter 5 RATIONAL NUMBERS: POSITIVE AND NEGATIVE DECIMALS

5.2 Rounding Decimal Numbers

Learning Objectives
1 Learn the rules for rounding decimals.
2 Round decimals to any given place.
3 Round money amounts to the nearest cent or nearest dollar.

Key Terms

Use the vocabulary terms listed below to complete each statement in exercises 1–2.

　　rounding　　　　　decimal places

1. _____ are the number of digits to the right of the decimal point.

2. When we "cut off" a number after a certain place value, we are _____ that number.

Objective 1 Learn the rules for rounding decimals.

For extra help, see page 335 of your text.

Objective 2 Round decimals to any given place.

Video Examples

Review these examples for Objective 2:

1. Round 16.87453 to the nearest thousandth.

   Step 1 Draw a "cut-off" line after the thousandths place.
   　16.874 | 53

   Step 2 Look only at the first digit you are cutting off. Ignore the other digits you are cutting off.
   　16.874 | 53   (Ignore the 3)

   Step 3 If the first digit you are cutting off is 5 or more, round up the part of the number you are keeping.
   　　16.874 | 53
   　$+ \ 0.001$
   　　16.875

   So, 16.87453 rounded to the nearest thousandth is 16.875. We can write $16.87453 \approx 16.875$.

Now Try:

1. Round 43.80290 to the nearest thousandth.

   _____

Name: Date:
Instructor: Section:

2. Round to the place indicated.

    d. 64.983 to the nearest tenth

    64.9 | 83

The first digit cut is 5 or more, so round up by adding 1 tenth to the part you are keeping.

$$\begin{array}{r} 1 \\ 64.9\ |\ 83 \\ +\ \ 0.1 \\ \hline 65.0 \end{array}$$

64.983 rounded to the nearest tenth is 65.0. We can write $64.983 \approx 65.0$. You must write the 0 in the tenths place to show that the number was rounded to the nearest tenth.

    a. 6.4387 to the nearest tenth

Step 1 Draw a cut-off line after the tenths place.
    6.4 | 387

Step 2 Look only at the 3.
    6.4 | 387   (Ignore the 8 and 7)

Step 3 The first digit is 4 or less, so the part you are keeping stays the same.
    6.4 | 387
    6.4

Rounding 6.4387 to the nearest tenth is 6.4. We can write $6.4387 \approx 6.4$.

2. Round to the place indicated.

    d. 0.649 to the nearest tenth

    _____

    a. 0.7976 to the nearest hundredth

    _____

Objective 2 Practice Exercises

For extra help, see Examples 1–2 on pages 335–337 of your text.

Round each number to the place indicated.

1. 489.84 to the nearest tenth   1. _____

2. 54.4029 to the nearest hundredth   2. _____

3. 989.98982 to the nearest thousandth   3. _____

Name: Date:
Instructor: Section:

Objective 3   Round money amounts to the nearest cent or nearest dollar.

Video Examples

Review these examples for Objective 3:

3. Round each money amount to the nearest cent.

   a. $6.5348

   Is $6.5348 closer to $6.53 or to $6.54?
   The first digit cut is 4 or less, so the part you are keeping stays the same.
   $6.53 | 48
   You pay $6.53.

   b. $0.895

   Is $0.895 closer to $0.89 or $0.90?
   $0.89 | 5
   The first digit cut is 5 or more, so round up.
   ```
         1
      $0.89 | 5
     + $0.01
      ─────
      $0.90
   ```
   You pay $0.90.

4. Round to the nearest dollar.

   c. $599.66

   Draw a cut-off line after the ones place.
   $599 | .66
   First digit cut is 5 or more, so round up by adding $1.
   ```
      $599 | .66
     +     1
      ─────
      $600
   ```
   So, $599.66 rounded to the nearest dollar is $600.

   e. $0.53

   $0 | .53
   First digit cut is 5 or more, so round up.
   $0.53 rounded to the nearest dollar is $1.

Now Try:

3. Round each money amount to the nearest cent.

   a. $2.0849

   _____

   b. $425.0954

   _____

4. Round to the nearest dollar.

   c. $880.83

   _____

   e. $0.61

Name: Date:
Instructor: Section:

Objective 3 Practice Exercises

For extra help, see Examples 3–4 on pages 338–339 of your text.

Round to the nearest dollar.

4. $28.39

4. _____

5. $11,839.73

5. _____

Round to the nearest cent.

6. $1028.6666

6. _____

Name: Date:
Instructor: Section:

Chapter 5 RATIONAL NUMBERS: POSITIVE AND NEGATIVE DECIMALS

5.3 Adding and Subtracting Signed Decimal Numbers

Learning Objectives
1. Add and subtract positive decimals.
2. Add and subtract negative decimals.
3. Estimate the answer when adding or subtracting decimals.

Key Terms

Use the vocabulary terms listed below to complete each statement in exercises 1–2.

estimating **front end rounding**

1. With _____, we round to the highest possible place.

2. Avoid common errors in working decimal problems by _____ the answer first.

Objective 1 **Add and subtract positive decimals.**

Video Examples

Review these examples for Objective 1:

1b. Find the sum.

$8.437 + 5.361 + 13.295$

Write the numbers vertically with decimal points lined up. Then add.

```
  1 1   1 1
    8 . 4 3 7
    5 . 3 6 1
 + 1 3 . 2 9 5
 ─────────────
   2 7 . 0 9 3
```

2b. Find the sum.

$12 + 9.36 + 3.754$

Write in zeros so that all the addends have three decimal places.

```
   1 2 . 0 0 0
       9 . 3 6 0
 +     3 . 7 5 4
 ─────────────
   2 5 . 1 1 4
```

Now Try:

1b. Find the sum.

$0.428 + 16.005 + 5.276$

_____

2b. Find the sum.

$0.631 + 999.3 + 14$

_____

Name: Date:
Instructor: Section:

4c. Find the difference.

16 less 7.54

Write a decimal point and two zeros after 16. Subtract as usual.

```
  16.00
-  7.54
  ─────
   8.46
```

4c. Find the difference.

1 less 0.499

_____

Objective 1 Practice Exercises

For extra help, see Examples 1–4 on pages 342–344 of your text.

Find each sum or difference.

1. 43.96 + 48.53

1. _____

2. 69.524 − 26.958

2. _____

3. 45.83 + 20.923 + 5.7

3. _____

Name: Date:
Instructor: Section:

Objective 2   Add and subtract negative decimals.

Video Examples

Review these examples for Objective 2:

5b. Find the sum.

$-4.36 + 0.973$

The addends have different signs. To begin, $|-4.36|$ is 4.36, and $|0.973|$ is 0.973. Then subtract the lesser absolute value from the greater.

$$\begin{array}{r} 4.3\,6\,0 \\ -\,0.9\,7\,3 \\ \hline 3.3\,8\,7 \end{array} \leftarrow \text{Write one 0}$$

$-4.36 + 0.973 = -3.387$

6a. Find the difference.

$5.4 - 13.89$

Rewrite subtractions as adding the opposite.
$5.4 - 13.89$
$5.4 + (-13.89)$

-13.89 has the greater absolute value and is negative, so the answer will be negative.

$5.4 + (-13.89) = -8.49$ $\quad\begin{array}{r} 1\,3.8\,9 \\ -\ \ 5.4\,0 \\ \hline 8.4\,9 \end{array}$

5a. Find the sum.

$-5.8 + (-21)$

Both addends are negative, so the sum will be negative. To begin, $|-5.8|$ is 5.8, and $|-21|$ is 21. Then add the absolute values.

$$\begin{array}{r} 5.8 \\ +\,2\,1.0 \\ \hline 2\,6.8 \end{array} \leftarrow \begin{array}{l}\text{Write in a decimal point} \\ \text{and one 0}\end{array}$$

$-5.8 + (-21) = -26.8$

Now Try:

5b. Find the sum.

$-13.57 + 2.984$

_____

6a. Find the difference.

$7.6 - 19.34$

_____

5a. Find the sum.

$-6.4 + (-32)$

_____

Name: Date:
Instructor: Section:

Objective 2 Practice Exercises

For extra help, see Examples 5–6 on pages 345–346 of your text.

Find each difference.

4. 18.1 − 84.6   4. _____

5. 87.6 − (−90.4)   5. _____

6. 1.71 − 12.68   6. _____

Objective 3 Estimate the answer when adding or subtracting decimals.

Video Examples

Review these examples for Objective 3:
7. Use front end rounding to round each number. Then add or subtract the rounded numbers to get an estimate answer. Finally, find the exact answer.

   a. Find the sum of 295.8 and 7.894.

   Estimate: *Exact:*
   300 ←Rounds to— 295.800
   + 8 ←Rounds to— + 7.894
   ───   ───────
   308   303.694

   The estimate goes out to the hundreds place and so does the exact answer. Therefore, the decimal point is probably in the correct place in the exact answer.

Now Try:
7. Use front end rounding to round each number. Then add or subtract the rounded numbers to get an estimate answer. Finally, find the exact answer.
   a. Find the sum of 5.74 and 8.107.

   _____

   _____

Name: Date:
Instructor: Section:

b. Subtract $16.95 from $78.23.

Estimate: *Exact:*
$80 ←Rounds to— $78.23
−20 ←Rounds to— − 16.95
$60 $61.28

b. Subtract $23.49 from $65.04.

_____

_____

Objective 3 Practice Exercises

For extra help, see Example 7 on pages 346–347 of your text.

First, use front end rounding and estimate each answer. Then add or subtract to find the exact answer.

7. 593.8
    27.93
   + 54.87

7.
Estimate_____

Exact _____

8. 7.69 − 20.85

8.
Estimate_____

Exact _____

9. −9.7 − 4.862

9.
Estimate_____

Exact _____

Name: Date:
Instructor: Section:

Chapter 5 RATIONAL NUMBERS: POSITIVE AND NEGATIVE DECIMALS

5.4 Multiplying Signed Decimal Numbers

Learning Objectives
1   Multiply positive and negative decimals.
2   Estimate the answer when multiplying decimals.

Key Terms

Use the vocabulary terms listed below to complete each statement in exercises 1–3.

    decimal places   **factor**   **product**

1. Each number in a multiplication problem is called a _____.

2. When multiplying decimal numbers, first multiply the numbers, then find the total number of _____ in both factors.

3. The answer to a multiplication problem is called the _____.

Objective 1   **Multiply positive and negative decimals.**

Video Examples

Review these examples for Objective 1:
1. Find the product of 6.23 and −5.4.

   Step 1 Multiply the numbers as if they were whole numbers.

   $$\begin{array}{r} 6.23 \\ \times\ \ 5.4 \\ \hline 2492 \\ 3115\ \ \\ \hline 33642 \end{array}$$

   Step 2 Count the total number of decimal places in both factors.

   $$\begin{array}{r} 6.23 \leftarrow 2\text{ decimal places} \\ \times\ \ 5.4 \leftarrow 1\text{ decimal place} \\ \hline 2492\ \ \ \ \ \text{3 total decimal places} \\ 3115\ \ \\ \hline 33642 \end{array}$$

   Step 3 Count over 3 places in the product and write the decimal point. Count from right to left.

Now Try:
1. Find the product of 2.51 and −4.3.

   _____

```
    6.23   ← 2 decimal places
×    5.4   ← 1 decimal place
   2492
  3115
  33.642         3 total decimal places
```

Step 4 The factors have different signs, so the product is negative. 6.23 times −5.4 is −33.642.

2. Find the product: $(-0.035)(-0.07)$.

   Start by multiply, then count decimal places.
   ```
       0.035  ← 3 decimal places
   ×   0.07   ← 2 decimal places
       245         5 total decimal places
   ```
   After multiplying, the answer has only three decimal places, but five are needed, so write two zeros on the left side of the answer. Then count over 5 places and write in the decimal point.
   ```
       0.035  ← 3 decimal places
   ×   0.07   ← 2 decimal places
      .00245       5 total decimal places
   ```
   The final product is 0.00245, which has five decimal places. The product is positive because the factors have the same sign.

2. Find the product $(-0.062)(-0.03)$.

   2. _____

Objective 1 Practice Exercises

For extra help, see Examples 1–2 on pages 352–353 of your text.

Find each product.

1.   −19.3
      × 4.7

    1. _____

2.   0.682
      × 3.9

    2. _____

3. $(-0.074)(-0.05)$

    3. _____

Name: Date:
Instructor: Section:

Objective 2 Estimate the answer when multiplying decimals.

Video Examples

Review this example for Objective 2:

3. First estimate the answer to $(74.56)(18.9)$ using front end rounding. Then find the exact answer.

Estimate:

```
   70  ← Rounds to
 × 20  ← Rounds to
 1400
```

Exact:

```
    7 4.5 6  ← 2 decimal places
  ×   1 8.9  ← 1 decimal place
    6 7 1 0 4    3 total decimal places
    5 9 6 4 8
    7 4 5 6
  1 4 0 9.1 8 4
```

Both the estimate and the exact answer go out to the thousands place, so the decimal point in 1409.184 is probably in the correct place.

Now Try:

3. First estimate the answer to $(22.43)(5.03)$ using front end rounding. Then find the exact answer.

_____

Objective 2 Practice Exercises

For extra help, see Example 3 on page 354 of your text.

First use front-end rounding and estimate the answer. Then multiply to find the exact answer.

4.   29.8
   × 3.4

4. Estimate_____

   Exact _____

5.   32.53
   × 23.26

5. Estimate_____

   Exact _____

6.   391.9
   × 7.74

6. Estimate_____

   Exact _____

Name: Date:
Instructor: Section:

Chapter 5 RATIONAL NUMBERS: POSITIVE AND NEGATIVE DECIMALS

5.5 Dividing Signed Decimal Numbers

Learning Objectives
1 Divide a decimal by an integer.
2 Divide a number by a decimal.
3 Estimate the answer when dividing decimals.
4 Use the order of operations with decimals.

Key Terms

Use the vocabulary terms listed below to complete each statement in exercises 1−4.

  repeating decimal   quotient   dividend   divisor

1. In a division problem, the number being divided is called the _____.

2. The number $0.8\overline{3}$ is an example of a _____.

3. The answer to a division problem is called the _____.

4. In the problem $6.39 \div 0.9$, 0.9 is called the _____.

Objective 1 Divide a decimal by an integer.

Video Examples

Review these examples for Objective 1:

1a. Find the quotient. Check the quotient by multiplying.

$24.48 \div (-4)$

First consider $24.48 \div 4$.
Rewrite the division problem.
Step 1 Write the decimal point in the quotient directly above the decimal point in the dividend.

$$4\overline{)24.48}$$

Step 2 Divide as if the numbers were whole numbers.

$$\begin{array}{r}6.12\\4\overline{)24.48}\end{array}$$

Check by multiplying the quotient times the divisor.

Now Try:

1a. Find the quotient. Check the quotient by multiplying.

$9.891 \div (-7)$

_____

Name: Date:
Instructor: Section:

$$\begin{array}{r}6.12\\\times4\\\hline 24.48\end{array}$$

Step 3 The quotient is −6.12 because the numbers have different signs.
$24.48 \div (-4) = -6.12$

2. Divide 1.35 by 4. Check the quotient by multiplying.

   Divide.

   $$\begin{array}{r}0.33\\4\overline{)1.35}\\\underline{1\,2}\\15\\\underline{12}\\3\end{array}$$

   Write a 0 after the 5 in the dividend so you can continue dividing. Keep writing more zeros in the dividend, if needed.

   $$\begin{array}{r}0.3375\\4\overline{)1.3500}\\\underline{1\,2}\\15\\\underline{12}\\30\\\underline{28}\\20\\\underline{20}\\0\end{array}$$
   Check: $\begin{array}{r}0.3375\\\times4\\\hline 1.3500\end{array}$

   The quotient is 0.3375.

2. Divide 1008.9 by 50. Check the quotient by multiplying.

   _____

3. Divide 8.87 by 9. Round the quotient to the nearest thousandth.

   Write extra zeros in the dividend so you can continue dividing.

   $$\begin{array}{r}0.9855\\9\overline{)8.8700}\\\underline{8\,1}\\77\\\underline{72}\\50\\\underline{45}\\50\\\underline{45}\\5\end{array}$$

3. Divide 302.24 by 18. Round the quotient to the nearest thousandth.

   _____

148 Copyright © 2018 Pearson Education, Inc.

Name: Date:
Instructor: Section:

Notice that the digit 5 in the answer is repeating. There are two ways to show that the answer is a repeating decimal that goes on forever.

   $0.9855\ldots$ or $0.98\overline{5}$

To round to thousandths, divide out one more place, to ten-thousandths.

   $8.87 \div 9 = 0.9855\ldots$ rounds to 0.986.

Check the answer by multiplying 0.986 by 9. Because 0.986 is a rounded answer, the check will not give exactly 8.87, but it should be very close.

   $(0.986)(9) = 8.874$

Objective 1 Practice Exercises

For extra help, see Examples 1–3 on pages 359–362 of your text.

Find each quotient. Round answers to the nearest thousandth, if necessary.

1. $5 \overline{) 34.8}$

1. _____

2. $-11 \overline{) 46.98}$

2. _____

3. $33 \overline{) 77.847}$

3. _____

Name: Date:
Instructor: Section:

Objective 2 Divide a number by a decimal.

Video Examples

Review this example for Objective 2:

4a. $\dfrac{41.2}{0.005}$

Move the decimal point in the divisor three places to the right so 0.005 becomes the whole number 5. Move the decimal point in the dividend the same number of places and write in two extra 0s.

$$5\overline{)41200.}= 8240.$$

Now Try:

4a. $0.0024\overline{)48.984}$

4a. _____

Objective 2 Practice Exercises

For extra help, see Example 4 on page 363 of your text.

Find each quotient. Round answers to the nearest thousandth, if necessary.

4. $0.9\overline{)3.4166}$ 4. _____

5. $3.4\overline{)436.05}$ 5. _____

6. $-0.07 \div (-0.00043)$ 6. _____

Name: Date:
Instructor: Section:

Objective 3 Estimate the answer when dividing decimals.

Video Examples

Review this example for Objective 3:
5. First use front end rounding to round each number and estimate the answer. Then divide to find the exact answer.
$$767.38 \div 3.7$$

Estimate:
$$4\overline{)800} = 200$$

Exact:
```
       27.4
37)7673.8
    74
    ---
    273
    259
    ---
    148
    148
    ---
      0
```

Notice that the estimate, which is in hundreds, is very different from the exact answer, which is in tens.
 Find the error and rework.
The exact answer is 207.4, which fits the estimate of 200.

Now Try:
5. First use front end rounding to round each number and estimate the answer. Then divide to find the exact answer.
$$185.22 \div 4.9$$

_____

Objective 3 Practice Exercises

For extra help, see Example 5 on page 364 of your text.

*Decide if each answer is **reasonable** or **unreasonable** by rounding the numbers and estimating the answer.*

7. $126.2 \div 11.2 = 11.268$ 7. _____

8. $31.5 \div 8.4 = 37.5$ 8. _____

9. $8695.15 \div 98.762 = 88.0415$ 9. _____

Name: Date:
Instructor: Section:

Objective 4   Use the order of operations with decimals.

Video Examples

Review these examples for Objective 4:

6. Use the order of operations to simplify each expression.

   b. $1.73 + (2.9 - 3.8)(6.5)$

   Work inside the parentheses.
   $1.73 + (-0.9)(6.5)$
   Multiply.
   $1.73 + (-5.85)$
   Add.
   -4.12

   c. $5.6^2 - 1.4 \div 7(2.4)$

   Apply the exponent.
   $31.36 - 1.4 \div 7(2.4)$
   Multiply and divide from left to right.
   $31.36 - 0.2(2.4)$
   Multiply.
   $31.36 - 0.48$
   Subtract last.
   30.88

Now Try:

6. Use the order of operations to simplify each expression.

   b. $4.06 \div 1.4 \times 7.8$

   _____

   c. $0.07 + 0.3(6.99 - 8)$

   _____

Objective 4 Practice Exercises

For extra help, see Example 6 on pages 364–365 of your text.

Use the order of operations to simplify each expression.

10. $(-3.1)^2 - 1.9 + 5.8$ 10. _____

11. $58.1 - (17.9 - 15.2) \times 1.8$ 11. _____

12. $9.1 - 0.07(2.1 \div 0.042)$ 12. _____

Name: Date:
Instructor: Section:

Chapter 5 RATIONAL NUMBERS: POSITIVE AND NEGATIVE DECIMALS

5.6 Fractions and Decimals

Learning Objectives
1   Write fractions as equivalent decimals.
2   Compare the size of fractions and decimals.

Key Terms

Use the vocabulary terms listed below to complete each statement in exercises 1–4.

    numerator   **denominator**   **mixed number**   **equivalent**

1. A fraction and a decimal that represent the same portion of a whole are _____.

2. The _____ of a fraction is the dividend.

3. The _____ of a fraction shows the number of equal parts in a whole.

4. A _____ consists of a whole number part and a fractional or decimal part.

Objective 1   Write fractions as equivalent decimals.

Video Examples

Review these examples for Objective 1:
1. Write the fraction as a decimal.

   a. $\dfrac{7}{8}$

   $\dfrac{7}{8}$ means $7 \div 8$. Write it as $8\overline{)7}$. Write extra zeros in the dividend so you can continue dividing until the remainder is zero.

   $$\begin{array}{r} 0.875 \\ 8\overline{)7.000} \\ \underline{6\ 4} \\ 60 \\ \underline{56} \\ 40 \\ \underline{40} \\ 0 \end{array}$$

   Therefore, $\dfrac{7}{8} = 0.875$

Now Try:
1. Write the fraction as a decimal.

   a. $\dfrac{1}{16}$

   _____

Name: Date:
Instructor: Section:

b. $3\frac{3}{16}$

One method is to divide 3 by 16 to get 0.1875 for the fraction part. Then add the whole number part to 0.1875.

$$\frac{3}{16} \rightarrow 16\overline{)3.0000} \rightarrow \begin{array}{r} 3.0000 \\ +\ 0.1875 \\ \hline 3.1875 \end{array}$$

$$\begin{array}{r} 0.1875 \\ 16\overline{)3.0000} \\ \underline{1\ 6} \\ 140 \\ \underline{128} \\ 120 \\ \underline{112} \\ 80 \\ \underline{80} \\ 0 \end{array}$$

So $3\frac{3}{16} = 3.1875$.

A second method is to first write $3\frac{3}{16}$ as an improper fraction and then divide numerator by denominator.

$$3\frac{3}{16} = \frac{51}{16}$$

$$\frac{51}{16} \rightarrow 51 \div 16 \rightarrow 16\overline{)51} \rightarrow 16\overline{)51.0000}$$

$$\begin{array}{r} 3.1875 \\ 16\overline{)51.0000} \\ \underline{48} \\ 30 \\ \underline{16} \\ 140 \\ \underline{128} \\ 120 \\ \underline{112} \\ 80 \\ \underline{80} \\ 0 \end{array}$$

So $3\frac{3}{16} = 3.1875$.

2. Write $\frac{5}{9}$ as a decimal and round to the nearest thousandth.

$\frac{5}{9}$ means $5 \div 9$. To round to thousandths, divide out one more place, to ten-thousandths.

b. $4\frac{3}{8}$

_____

2. Write $\frac{5}{12}$ as a decimal and round to the nearest thousandth.

_____

Name: Date:
Instructor: Section:

$$\frac{5}{9} \to 5 \div 9 \to 9\overline{)5} \to 9\overline{)5.0000}^{\,0.5555}$$

$$\phantom{9\overline{)5.0000}}\underline{45}$$
$$\phantom{9\overline{)5.00}}50$$
$$\phantom{9\overline{)5.000}}\underline{45}$$
$$\phantom{9\overline{)5.000}}50$$
$$\phantom{9\overline{)5.0000}}\underline{45}$$
$$\phantom{9\overline{)5.0000}}50$$
$$\phantom{9\overline{)5.00000}}\underline{45}$$
$$\phantom{9\overline{)5.00000}}5$$

Written as a repeating decimal, $\frac{5}{9} = 0.\overline{5}$.

Rounded to the nearest thousandth, $\frac{5}{9} = 0.556$.

Objective 1 Practice Exercises

For extra help, see Examples 1–2 on pages 372–374 of your text.

Write each fraction or mixed number as a decimal. Round to the nearest thousandth, if necessary.

1. $\frac{7}{8}$

1. _____

2. $4\frac{1}{9}$

2. _____

3. $19\frac{17}{24}$

3. _____

Name: Date:
Instructor: Section:

Objective 2   Compare the size of fractions and decimals.
Video Examples
Review these examples for Objective 2:

3. Write in <, >, or = in the blank between each pair of numbers.

   a. 0.735 ____ 0.75

   Because 0.735 is to the left of 0.75, use the < symbol.
   0.735 is less than 0.75 can be written as
   $0.735 < 0.75$

   b. $\frac{3}{5}$ ____ 0.6

   $\frac{3}{5}$ and 0.6 are the same point on the number line.
   They are equivalent.
   $\frac{3}{5} = 0.6$

   c. $0.\overline{3}$ ____ 0.3

   0.3 is to the right of $0.\overline{3}$ (which is actually 0.333...), so use the > symbol.
   $0.\overline{3}$ is greater than 0.3 can be written as
   $0.\overline{3} > 0.3$

4a. Write the group of numbers in order, from least to greatest.

   0.58   0.575   0.5816

   Write zeros to the right of 0.58 and 0.575, so they also have four decimal places. Then find the least and greatest number of ten-thousandths.
   $0.58 = 0.5800 = 5800$ ten-thousandths middle
   $0.575 = 0.5750 = 5750$ ten-thousandths least
   $\quad\quad = 0.5816 = 5816$ ten-thousandths greatest
   From least to greatest, the correct order is
   0.575   0.58   0.5816

Now Try:

3. Write in <, >, or = in the blank between each pair of numbers.

   a. 0.85 ____ 0.7895

   _____

   b. $\frac{5}{16}$ ____ 0.3125

   _____

   c. $\frac{3}{7}$ ____ $0.\overline{4}$

   _____

4a. Write the group of numbers in order, from least to greatest.

   0.3   0.307   0.3057

   _____

Name: Date:
Instructor: Section:

Objective 2 Practice Exercises

For extra help, see Examples 3–4 on pages 374–375 of your text.

Write < or > to make a true statement.

4. $\dfrac{5}{6}$ ___ 0.83   4. _____

Arrange in order from least to greatest.

5. $\dfrac{3}{11}, \dfrac{1}{3}, 0.29$ 5. _____

6. $1.085, 1\dfrac{5}{11}, 1\dfrac{7}{20}$ 6. _____

Name: Date:
Instructor: Section:

Chapter 5 RATIONAL NUMBERS: POSITIVE AND NEGATIVE DECIMALS

5.7 Problem Solving with Statistics: Mean, Median, and Mode

Learning Objectives
1. Find the mean of a list of numbers.
2. Find a weighted mean.
3. Find the median.
4. Find the mode.

Key Terms

Use the vocabulary terms listed below to complete each statement in exercises 1–4.

   mean   weighted mean   median   mode

1. The _____ is the value that occurs most often in a group of values.

2. A mean calculated so that each value is multiplied by its frequency is called a _____.

3. The sum of all the values in a data set divided by the number of values in the data set is called the _____.

4. The middle number in a group of values that are listed from smallest to largest is called the _____.

Objective 1 Find the mean of a list of numbers.

Video Examples

Review this example for Objective 1:
2. Find the mean of 31, 37, 44, 51, 52, 74, 69, and 83.

   Find the mean (rounded to the nearest tenth).
   $$\text{mean} = \frac{31+37+44+51+52+74+69+83}{8}$$
   $$\text{mean} = \frac{441}{8}$$
   $$\text{mean} \approx 55.1$$
   The mean is 55.1.

Now Try:
2. Find the mean of 40.1, 32.8, 82.5, 51.2, 88.3, 31.7, 43.7, and 51.2.

   _____

Name: Date:
Instructor: Section:

Objective 1 Practice Exercises

For extra help, see Examples 1–2 on page 380 of your text.

Find the mean for each list of numbers. Round to the nearest tenth, if necessary.

1.   39, 50, 59, 61, 69, 73, 51, 80   1. _____

2.   62.7, 59.6, 71.2, 65.8, 63.1   2. _____

3.   216, 245, 268, 268, 280, 291, 304, 313   3. _____

Objective 2   Find a weighted mean.

Video Examples

Review this example for Objective 2:

3. Use the following table to find the weighted mean.

Value	Frequency
13	4
12	2
19	5
15	3
21	1
27	5

To find the mean, multiply the value by the frequency. Then add the products. Next, add the numbers in the frequency column to find the total number of values.

Now Try:

3. Use the following table to find the weighted mean.

Value	Frequency
35	1
36	2
39	5
40	4
42	3
43	5

_____

Name: Date:
Instructor: Section:

Value	Frequency	Product
13	4	$(13 \cdot 4) = 52$
12	2	$(12 \cdot 2) = 24$
19	5	$(19 \cdot 5) = 95$
15	3	$(15 \cdot 3) = 45$
21	1	$(21 \cdot 1) = 21$
27	5	$(27 \cdot 5) = 135$
Totals	20	372

Finally, divide the totals. Round to the nearest tenth.

$$\text{mean} = \frac{372}{20} = 18.6$$

The mean is 18.6.

Objective 2 Practice Exercises

For extra help, see Examples 3–4 on pages 381–382 of your text.

Find the weighted mean for each list of numbers. Round to the nearest tenth, if necessary.

4.
Value	Frequency
17	4
12	5
15	3
19	1

4. _____

5.
Value	Frequency
1	2
2	3
4	5
5	7
6	4
7	2
8	1
9	1

5. _____

Name: Date:
Instructor: Section:

Find the grade point average for this student. Assume A = 4, B = 3, C = 2, D = 1, F = 0.

6.
Units	Grade
5	B
4	C
3	B
2	C
2	C

6. _____

Objective 3   Find the median.

Video Examples

Review these examples for Objective 3:

5. Find the median for the list of values.
   21, 32, 27, 23, 25, 29, 22

   First arrange the numbers in numerical order from least to greatest.
   21, 22, 23, 25, 27, 29, 32
   Next, find the middle number in the list.
   21, 22, 23,   25,   27, 29, 32
   Three are below.   Three are above.
               Middle number
   The median value is 25.

6. Find the median for the list of values.
   389, 464, 521, 610, 654, 672, 682, 712

   First arrange the numbers in numerical order from least to greatest. Then find the middle two numbers.
   389, 464, 521, 610, 654, 672, 682, 712
                Middle two numbers
   The median value is the mean of the two middle numbers.
   $$\text{median} = \frac{610+654}{2} = \frac{1264}{2} = 632$$
   The median value is 632.

Now Try:

5. Find the median for the list of values.
   18, 12, 11, 19, 26

   _____

6. Find the median for the list of values.
   0.02, 0.04, 0.12, 0.08

   _____

Name: Date:
Instructor: Section:

Objective 3 Practice Exercises

For extra help, see Examples 5–6 on pages 382–383 of your text.

Find the median for each list of numbers.

7. 200, 215, 226, 238, 250, 283

7. _____

8. 43, 69, 108, 32, 51, 49, 83, 57, 64

8. _____

9. 200, 195, 302, 284, 256, 237, 239, 240

9. _____

Objective 4   Find the mode.

Video Examples

Review this example for Objective 4:
7b. Find the mode for the list of numbers.

964, 987, 973, 987, 921, 921, 975

Because both 987 and 921 occur twice, each is a mode.

Now Try:
7b. Find the mode for the list of numbers.
16, 13, 21, 16, 18, 11, 13, 15, 14

_____

Objective 4 Practice Exercises

For extra help, see Example 7 on page 383 of your text.

Find the mode for each list of numbers.

10. 4, 9, 3, 4, 7, 3, 2, 3, 9

10. _____

11. 37, 24, 35, 35, 24, 38, 39, 28, 27, 39

11. _____

12. 172.6, 199.7, 182.4, 167.1, 172.6, 183.4, 187.6

12. _____

Name: Date:
Instructor: Section:

Chapter 5 RATIONAL NUMBERS: POSITIVE AND NEGATIVE DECIMALS

5.8 Geometry Applications: Pythagorean Theorem and Square Roots

Learning Objectives
1. Find square roots using the square root key on a calculator.
2. Find the unknown length in a right triangle.
3. Solve application problems involving right triangles.

Key Terms

Use the vocabulary terms listed below to complete each statement in exercises 1–4.

    hypotenuse   **legs**   **right triangle**   **square root**

1. A triangle with a 90° angle is called a _____.

2. The side opposite the right angle in a right triangle is called the _____ of the triangle.

3. A positive _____ of a positive number is one of two identical positive factors of the number.

4. The two sides of the right angle in a right triangle are called the _____ of the triangle.

Objective 1   Find square roots using the square root key on a calculator.

Video Examples

Review these examples for Objective 1:	Now Try:

1. Use a calculator to find each square root. Round answers to the nearest hundredth.
 	
   a. $\sqrt{45}$
 	
   Calculator shows 6.708203932; round to 6.71.
 	
   b. $\sqrt{86}$
 	
   Calculator shows 9.273618495; round to 9.27.
	1. Use a calculator to find each square root. Round answers to the nearest hundredth.
	
	   a. $\sqrt{20}$
	
	   _____
	
	   b. $\sqrt{92}$
	
	   _____

Name: Date:
Instructor: Section:

Objective 1 Practice Exercises

For extra help, see Example 1 on page 388 of your text.

Find each square root. Use a calculator with a square root key. Round the answer to the nearest thousandth, if necessary.

1. $\sqrt{17}$ 1. _____

2. $\sqrt{75}$ 2. _____

3. $\sqrt{102}$ 3. _____

Objective 2 Find the unknown length in a right triangle.

Video Examples

Review these examples for Objective 2:
2. Find the unknown length in each right triangle. Round answers to the nearest tenth if necessary.

   a.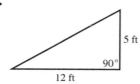

   The unknown length is the side opposite the right angle, which is the hypotenuse. Use the formula for finding the hypotenuse.

   $$\text{hypotenuse} = \sqrt{(\text{leg})^2 + (\text{leg})^2}$$
   $$\text{hypotenuse} = \sqrt{(5)^2 + (12)^2}$$
   $$= \sqrt{25 + 144}$$
   $$= \sqrt{169}$$
   $$= 13$$

   The hypotenuse is 13 ft long.

   b.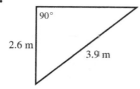

   We do know the length of the hypotenuse (3.9 m), so it is the length of one of the legs that

Now Try:
2. Find the unknown length in each right triangle. Round answers to the nearest tenth if necessary.

   a.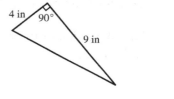
   _____

   b.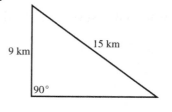
   _____

164 Copyright © 2018 Pearson Education, Inc.

Name: Date:
Instructor: Section:

is unknown. Use the formula for finding the leg.

$$\text{leg} = \sqrt{(\text{hypotenuse})^2 - (\text{leg})^2}$$
$$\text{leg} = \sqrt{(3.9)^2 - (2.6)^2}$$
$$= \sqrt{15.21 - 6.76}$$
$$= \sqrt{8.45}$$
$$\approx 2.9$$

The length of the leg is approximately 2.9 m.

Objective 2 Practice Exercises

For extra help, see Example 2 on page 389 of your text.

Find the unknown length in each right triangle. Use a calculator with a square root key. Round the answer to the nearest tenth, if necessary.

4.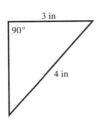

4. _____

5.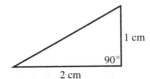

5. _____

6.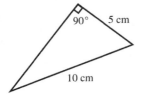

6. _____

Name: Date:
Instructor: Section:

Objective 3 Solve application problems involving right triangles.

Video Examples

Review this example for Objective 3:

3. The base of a ladder is located 7 feet from a building. The ladder reaches 24 feet up the building. How long is the ladder? Round to the nearest tenth of a foot if necessary.

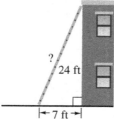

Now Try:

3. Find the unknown length in this roof plan. Round to the nearest tenth of a foot if necessary.

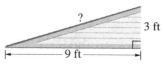

_____

A right triangle is formed. The unknown side is the hypotenuse.

$$\text{hypotenuse} = \sqrt{(\text{leg})^2 + (\text{leg})^2}$$
$$\text{hypotenuse} = \sqrt{(7)^2 + (24)^2}$$
$$= \sqrt{49 + 576}$$
$$= \sqrt{625}$$
$$= 25$$

The length of the ladder is 25 ft.

Objective 3 Practice Exercises

For extra help, see Example 3 on page 390 of your text.

Solve each application problem. Draw a diagram if one is not provided. Use a calculator with a square root key. Round the answer to the nearest tenth, if necessary.

7. Find the length of the loading ramp. 7. _____

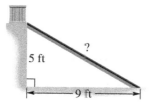

166 Copyright © 2018 Pearson Education, Inc.

Name: Date:
Instructor: Section:

8. A kite is flying on 50 feet of string. If the horizontal distance of the kite from the person flying it is 40 feet, how far off the ground is the kite?

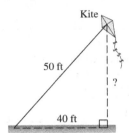

8. _____

9. The base of a 17-ft ladder is located 15 ft from a building. How high up on the building will the ladder reach?

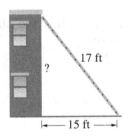

9. _____

Name: Date:
Instructor: Section:

Chapter 5 RATIONAL NUMBERS: POSITIVE AND NEGATIVE DECIMALS

5.9 Problem Solving: Equations Containing Decimals

Learning Objectives
1   Solve equations containing decimals using the addition property of equality.
2   Solve equations containing decimals using the division property of equality.
3   Solve equations containing decimals using both properties of equality.
4   Solve application problems involving equations with decimals.

Key Terms

Use the vocabulary terms listed below to complete each statement in exercises 1−2.

addition property of equality

division property of equality

1. The _____ states that both sides of an equation may be divided by the same nonzero number and it will still be balanced.

2. The _____ states that the same number may be added to both sides of an equation and the equation will still be balanced.

Objective 1 Solve equations containing decimals using the addition property of equality.

Video Examples

Review these examples for Objective 1:
1. Solve each equation and check each solution.

   a. $w + 3.8 = -0.7$

   Use the addition property to "get rid of" the 3.8 on the left side by adding its opposite, −3.8.

   $$w + 3.8 = -0.7$$
   $$\underline{-3.8 \quad -3.8}$$
   $$w + 0 = -4.5$$
   $$w = -4.5$$

   The solution is −4.5.

   Check. Go back to the original equation and replace w with −4.5.
   $$w + 3.8 = -0.7$$
   $$-4.5 + 3.8 = -0.7$$
   $$-0.7 = -0.7$$
   So −4.5 is the correct solution.

Now Try:
1. Solve each equation and check each solution.
   a. $w + 4.2 = -9.7$

   _____

168 Copyright © 2018 Pearson Education, Inc.

Name: Date:
Instructor: Section:

b. $8 = -5.6 + x$

To get x by itself on the right side of the equal sign, add 5.6 to both sides.

$$8 = -5.6 + x$$
$$\underline{+5.6 \quad +5.6}$$
$$13.6 = 0 + x$$
$$13.6 = x$$

The solution is 13.6.

Check. Go back to the original equation and replace x with 13.6.
$$8 = -5.6 + x$$
$$8 = -5.6 + 13.6$$
$$8 = 8$$
So 13.6 is the correct solution.

b. $11 = -8.6 + x$

Objective 1 Practice Exercises

For extra help, see Example 1 on page 395 of your text.

Solve each equation and check each solution.

1. $-30.4 + n = -35$

   1. _____

2. $-7.1 = 0.32 + m$

   2. _____

3. $8.7 + h = 16.4$

   3. _____

Name: Date:
Instructor: Section:

Objective 2 Solve equations containing decimals using the division property of equality.

Video Examples

Review these examples for Objective 2:
2. Solve each equation and check each solution.

   a. $6x = 25.5$

   To undo the multiplication by 6, divide both sides by 6.
   $$6x = 25.5$$
   $$\frac{6 \cdot x}{6} = \frac{25.5}{6}$$
   $$\frac{\cancel{6} \cdot x}{\cancel{6}} = 4.25$$
   $$x = 4.25$$
   The solution is 4.25.

   Check. Go back to the original equation and replace x with 4.25.
   $$6x = 25.5$$
   $$6(4.25) = 25.5$$
   $$25.5 = 25.5$$
   So 4.25 is the correct solution.

   b. $-10.5 = 1.4t$

   Divide both sides by the coefficient of the variable term, 1.4.
   $$-10.5 = 1.4t$$
   $$\frac{-10.5}{1.4} = \frac{\cancel{1.4}\, t}{\cancel{1.4}}$$
   $$-7.5 = t$$
   The solution is -7.5.

   Check. Go back to the original equation and replace t with -7.5.
   $$-10.5 = 1.4t$$
   $$-10.5 = 1.4(-7.5)$$
   $$-10.5 = -10.5$$
   So -7.5 is the correct solution.

Now Try:
2. Solve each equation and check each solution.
   a. $8x = 24.4$

   _____

   b. $-54.4 = 6.4t$

   _____

Name: Date:
Instructor: Section:

Objective 2 Practice Exercises

For extra help, see Example 2 on page 396 of your text.

Solve each equation and check each solution.

4. $-3y = -0.96$ 4. _____

5. $2.7r = 6.75$ 5. _____

6. $87.6 = -12r$ 6. _____

Objective 3 Solve equations containing decimals using both properties of equality.

Video Examples

Review this example for Objective 3:
3b. Solve the equation and check the solution.

$$7x - 0.87 = 3x + 2.49$$

Use the addition property to "get rid of" $3x$ on the right side by adding its opposite, $-3x$, to both sides.

$$\begin{aligned} 7x - 0.87 &= 3x + 2.49 \\ -3x & -3x \\ \hline 4x - 0.87 &= 0 + 2.49 \\ 4x + (-0.87) &= 2.49 \\ +0.87 & +0.87 \\ \hline 4x + 0 &= 3.36 \end{aligned}$$

$$\frac{\cancel{4}x}{\cancel{4}} = \frac{3.36}{4}$$

$$x = 0.84$$

The solution is 0.84.

Now Try:
3b. Solve the equation and check the solution.
$9x - 1.46 = 3x + 1.72$

_____

Name: Date:
Instructor: Section:

Check. Go back to the original equation and replace x with 0.84.
$$7x - 0.87 = 3x + 2.49$$
$$7(0.84) - 0.87 = 3(0.84) + 2.49$$
$$5.88 - 0.87 = 2.52 + 2.49$$
$$5.01 = 5.01$$
So 0.84 is the correct solution.

Objective 3 Practice Exercises

For extra help, see Example 3 on page 397 of your text.

Solve each equation and check each solution.

7. $2.2 = 0.5y + 9.7$ 7. _____

8. $-11.2 - 0.7p = 0.9p + 5.6$ 8. _____

9. $0.6x + 4.98 = x - 6.78$ 9. _____

Name: Date:
Instructor: Section:

Objective 4   Solve application problems involving equations with decimals.

Video Examples

Review this example for Objective 4:

Now Try:

4. When using a particular calling card, the cost is $1.15 per minute plus a $0.75 service charge per call. If Julie was billed $20.30 for one call, how long did the call last?

Step 1 Read the problem. It is about the cost of the call.

Step 2 Assign a variable. Let m be the number of minutes for the call.

Step 3 Write an equation.

Cost per minute	Number of minutes		Service charge		Total cost
1.15	· m	+	0.75	=	20.30

Step 4 Solve the equation.

$$1.15m + 0.75 = 20.30$$
$$\underline{-0.75 \quad -0.75}$$
$$1.15m + 0 = 19.55$$
$$\frac{1.15m}{1.15} = \frac{19.55}{1.15}$$
$$m = 17$$

Step 5 State the answer. The call was 17 min.

Step 6 Check the solution.
    $1.15 per minute times 17 minutes is $19.55
    $19.55 plus $0.75 service charge is $20.30.
The answer checks.

4. A telephone company charges $4.25 plus an additional $0.07 per minute for long distance calls. How many minutes did Sheila talk if her bill for a call was $6.49?

                _____

Objective 4 Practice Exercises

For extra help, see Example 4 on page 398 of your text.

Solve each application problem using the six problem-solving steps.

10. An air compressor can be rented for $38.95 for the first three hours, and $8.50 for each additional hour. William's rental charge was $64.45. How many hours did he rent the compressor?

    10. _____

Name: Date:
Instructor: Section:

11. Most adult medication doses are for a person weighing 150 pounds. For a 45-pound child, the adult dose should be multiplied by 0.3. If the child's dose of a decongestant is 9 milligrams, what is the adult dose?

11. _____

12. The bill for a three-night hotel stay was $371.43. This included $42.15 for room service and $35.28 room tax. What was the room rate per night?

12. _____

Name: Date:
Instructor: Section:

Chapter 5 RATIONAL NUMBERS: POSITIVE AND NEGATIVE DECIMALS

5.10   Geometry Applications: Circles, Cylinders, and Surface Area

Learning Objectives
1   Find the radius and diameter of a circle.
2   Find the circumference of a circle.
3   Find the area of a circle.
4   Find the volume of a cylinder.
5   Find the surface area of a rectangular solid.
6   Find the surface area of a cylinder.

Key Terms

Use the vocabulary terms listed below to complete each statement in exercises 1−6.

circle   radius   diameter   circumference   π (pi)

surface area

1.   The _____ is the distance from the center of a circle to any point on the circle.

2.   The _____ of a circle is the distance around the circle.

3.   A figure whose points lie the same distance from a fixed center point is called a _____.

4.   The ratio of the circumference to the diameter of any circle equals _____.

5.   The area on the surface of a three-dimensional object is called its _____.

6.   The _____ of a circle is a segment connecting two points on a circle and passing through the center.

Name: Date:
Instructor: Section:

Objective 1 Find the radius and diameter of a circle.

Video Examples

Review this example for Objective 1:
1b. Find the unknown length in the circle.

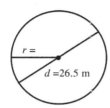

The radius is half the diameter.
$$r = \frac{d}{2}$$
$$r = \frac{26.5 \text{ m}}{2}$$
$$r = 13.25 \text{ m}$$

Now Try:
1b. Find the unknown length in the circle.

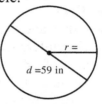

_____

Objective 1 Practice Exercises

For extra help, see Example 1 on pages 401–402 of your text.

Find the diameter or radius in each circle.

1. The diameter of a circle is 8 feet. Find its radius.

1. _____

2. The radius of a circle is 2.7 centimeters. Find its diameter.

2. _____

3. The diameter of a circle is $12\frac{1}{2}$ yards. Find its radius.

3. _____

Name: Date:
Instructor: Section:

Objective 2 Find the circumference of a circle.

Video Examples

Review this example for Objective 2:
2a. Find the circumference of the circle. Use 3.14 as the approximate value for π. Round answer to the nearest tenth.

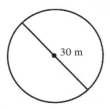

The diameter is 30 m, so use the formula with d in it.
$C = \pi \cdot d$
$C \approx 3.14 \cdot 30$ m
$C \approx 94.2$ m

Now Try:
2a. Find the circumference of the circle. Use 3.14 as the approximate value for π. Round answer to the nearest tenth.

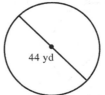

_____

Objective 2 Practice Exercises

For extra help, see Example 2 on page 403 of your text.

Find the circumference of each circle. Use 3.14 as an approximation for π. Round each answer to the nearest tenth.

4. A circle with a diameter of $4\frac{3}{4}$ inches   4. _____

5. A circle with a radius of 4.5 yards   5. _____

6. A circle with a radius of 8 cm   6. _____

Name: Date:
Instructor: Section:

Objective 3 Find the area of a circle.

Video Examples

Review this example for Objective 3:

3b. Find the area of the circle. Use 3.14 as the approximate value for π. Round your answer to the nearest tenth.

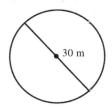

Now Try:

3b. Find the area of the circle. Use 3.14 as the approximate value for π. Round your answer to the nearest tenth.

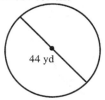

_____

To use the area formula, you need to know the radius (r). In this circle, the diameter is 30 m. First find the radius.

$r = \dfrac{d}{2}$

$r = \dfrac{30 \text{ m}}{2} = 15 \text{ m}$

Now find the area.

$A = \pi \cdot r \cdot r$

$A \approx 3.14 \cdot 15 \text{ m} \cdot 15 \text{ m}$

$A \approx 706.5 \text{ m}^2$

Objective 3 Practice Exercises

For extra help, see Examples 3–6 on pages 404–406 of your text.

Find the area of each circle. Use 3.14 as an approximation for π. Round each answer to the nearest tenth.

7. A circle with diameter of $5\dfrac{1}{3}$ yards

7. _____

8. A circle with radius of 9.8 centimeters

8. _____

Name: Date:
Instructor: Section:

9.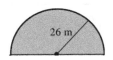

9. _____

Objective 4 Find the volume of a cylinder.

Video Examples

Review this example for Objective 4:

7a. Find the volume of the cylinder. Use 3.14 as the approximate value of π. Round your answer to the nearest tenth if necessary.

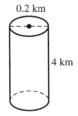

The diameter is 0.2 km, so the radius is $0.2 \text{ km} \div 2 = 0.1 \text{ km}$. The height is 4 km.

$V = \pi \cdot r \cdot r \cdot h$

$V \approx 3.14 \cdot 0.1 \text{ km} \cdot 0.1 \text{ km} \cdot 4 \text{ km}$

$V \approx 0.1 \text{ km}^3$

Now Try:

7a. Find the volume of the cylinder. Use 3.14 as the approximate value of π. Round your answer to the nearest tenth if necessary.

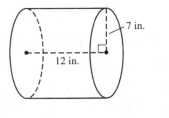

_____

Objective 4 Practice Exercises

For extra help, see Example 7 on page 406 of your text.

Find the volume of each figure. Use 3.14 as an approximation for π. Round answers to the nearest tenth, if necessary.

10.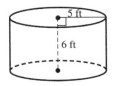

10. _____

Name: Date:
Instructor: Section:

11.   18 cm

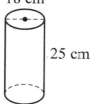

25 cm

11. _____

12.   A soup can, diameter 6 inches and height 8 inches

12. _____

Objective 5   Find the surface area of a rectangular solid.

Video Examples

Review this example for Objective 5:
8.   Find the volume and surface area of the figure.

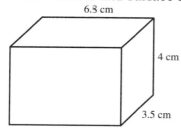

First find the volume.
$V = lwh$

$V = 6.8 \text{ cm} \cdot 3.5 \text{ cm} \cdot 4 \text{ cm}$

$V = 95.2 \text{ cm}^3$

Now find the surface area.
$S = 2lw + 2lh + 2wh$

$S = (2 \cdot 6.8 \text{ cm} \cdot 3.5 \text{ cm}) + (2 \cdot 6.8 \text{ cm} \cdot 4 \text{ cm})$
$\quad + (2 \cdot 3.5 \text{ cm} \cdot 4 \text{ cm})$

$S = 47.6 \text{ cm}^2 + 54.4 \text{ cm}^2 + 28 \text{ cm}^2$

$S = 130 \text{ cm}^2$

Now Try:
8.   Find the volume and surface area of the figure.

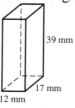

_____

Name: Date:
Instructor: Section:

Objective 5 Practice Exercises

For extra help, see Example 8 on page 407 of your text.

Find the surface area of each rectangular solid. Round your answers to the nearest tenth.

13.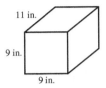

13. _____

14.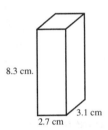

14. _____

15.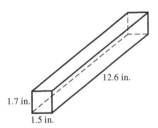

15. _____

Objective 6 Find the surface area of a cylinder.

Video Examples

Review this example for Objective 6:
9. Find the volume and surface area of the figure. Use 3.14 as the approximate value of π. Round your answer to the nearest tenth if necessary.

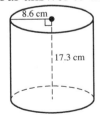

Now Try:
9. Find the volume and surface area of the figure. Use 3.14 as the approximate value of π. Round your answer to the nearest tenth if necessary.

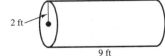

Name: Date:
Instructor: Section:

First find the volume.
$V = \pi r^2 h$
$V \approx 3.14 \cdot 8.6 \text{ cm} \cdot 8.6 \text{ cm} \cdot 17.3 \text{ cm}$
$V \approx 4017.7 \text{ cm}^3$

Now find the surface area.
$S = 2\pi rh + 2\pi r^2$
$S \approx (2 \cdot 3.14 \cdot 8.6 \text{ cm} \cdot 17.3 \text{ cm})$
$\quad + (2 \cdot 3.14 \cdot 8.6 \text{ cm} \cdot 8.6 \text{ cm})$
$S \approx 934.3384 \text{ cm}^2 + 464.4688 \text{ cm}^2$
$S \approx 1398.8 \text{ cm}^2$

Objective 6 Practice Exercises

For extra help, see Example 9 on page 408 of your text.

Find the surface area of each cylinder. Use 3.14 as the approximate value for π. Round your answers to the nearest tenth.

16.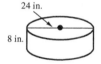

16. _____

17.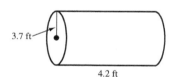

17. _____

18.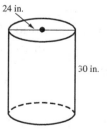

18. _____

Name: Date:
Instructor: Section:

Chapter 6 RATIO, PROPORTION, AND LINE/ANGLE/TRIANGLE RELATIONSHIPS

6.1 Ratios

Learning Objectives	
1	Write ratios as fractions.
2	Solve ratio problems involving decimals or mixed numbers.
3	Solve ratio problems after converting units.

Key Terms

Use the vocabulary terms listed below to complete each statement in exercises 1–2.

denominator   **numerator**   **ratio**

1. A _____ can be used to compare two measurements with the same type of units.

2. When writing the ratio to compare the width of a room to its height, the width goes in the _____ and the height goes in the _____.

Objective 1   Write ratios as fractions.

Video Examples

Review this example for Objective 1:

2a. Write the ratio in lowest terms.

   32 inches of snow to 8 inches of rain

   The ratio is $\frac{32 \text{ inches}}{8 \text{ inches}}$. Divide out the common units. Then write this ratio in lowest terms by dividing the numerator and denominator by 8.

   $\frac{32 \text{ inches}}{8 \text{ inches}} = \frac{32}{8} = \frac{32 \div 8}{8 \div 8} = \frac{4}{1}$

   So, the ratio of 32 inches of snow to 8 inches of rain is 4 to 1, or $\frac{4}{1}$. For every 4 inches of snow there is 1 inch of rain.

Now Try:

2a. Write the ratio in lowest terms.

   16 seconds to 24 seconds

   _____

Name: Date:
Instructor: Section:

Objective 1 Practice Exercises

For extra help, see Examples 1–2 on pages 432–434 of your text.

Write each ratio as a fraction in lowest terms.

1. 125 cents to 95 cents

    1. _____

2. 80 miles to 30 miles

    2. _____

3. 5 men to 20 men

    3. _____

Objective 2   Solve ratio problems involving decimals or mixed numbers.

Video Examples

Review this example for Objective 2:

4b. Write the ratio as a comparison of whole numbers in lowest terms.

$$7\frac{1}{5} \text{ to } 2\frac{1}{4}$$

Write the ratio as $\dfrac{7\frac{1}{5}}{2\frac{1}{4}}$. Then write $7\frac{1}{5}$ and $2\frac{1}{4}$ as improper fractions.

$$7\frac{1}{5} = \frac{36}{5} \quad \text{and} \quad 2\frac{1}{4} = \frac{9}{4}$$

The ratio is shown here.

$$\frac{7\frac{1}{5}}{2\frac{1}{4}} = \frac{\frac{36}{5}}{\frac{9}{4}}$$

Rewrite as a division problem in horizontal format, using the "÷" symbol for division. Then multiply by the reciprocal of the divisor.

$$\frac{36}{5} \div \frac{9}{4} = \frac{\cancel{36}^{4}}{5} \cdot \frac{4}{\cancel{9}_{1}} = \frac{16}{5}$$

Now Try:

4b. Write the ratio as a comparison of whole numbers in lowest terms.

$$3\frac{2}{3} \text{ to } 2\frac{5}{6}$$

_____

184 Copyright © 2018 Pearson Education, Inc.

Name: Date:
Instructor: Section:

Objective 2 Practice Exercises

For extra help, see Examples 3–4 on pages 434–435 of your text.

Write each ratio as a fraction in lowest terms.

4. $4\frac{1}{8}$ to 3

4. _____

5. 11 to $2\frac{4}{9}$

5. _____

Solve. Write each ratio as a fraction in lowest terms.

6. One car has a $15\frac{1}{2}$ gallon gas tank while another has a 22 gallon gas tank. Find the ratio of the amount the first tank holds to the amount the second tank holds.

6. _____

Objective 3   Solve ratio problems after converting units.

Video Examples

Review this example for Objective 3:

5a. Write the ratio of length of a board 3 ft long to the length of another board that is 42 inches long.

First, express 3 ft in inches. Because 1 ft has 12 in., 3 ft is
$3 \cdot 12$ in. $= 36$ in.
The ratio of the lengths is
$$\frac{3 \text{ ft}}{42 \text{ in.}} = \frac{36 \text{ in.}}{42 \text{ in.}} = \frac{36}{42}$$
Write the ratio in lowest terms.
$$\frac{36}{42} = \frac{36 \div 6}{42 \div 6} = \frac{6}{7}$$
The shorter board is $\frac{6}{7}$ the length of the longer board.

Now Try:

5a. Write the ratio of 20 days to 4 weeks.

_____

Name: Date:
Instructor: Section:

Objective 3 Practice Exercises

For extra help, see Example 5 on page 436 of your text.

Write each ratio as a fraction in lowest terms. Be sure to convert units as necessary.

7. 4 days to 2 weeks

7. _____

8. 6 yards to 10 feet

8. _____

9. 40 ounces to 3 pounds

9. _____

Name: Date:
Instructor: Section:

Chapter 6 RATIO, PROPORTION, AND LINE/ANGLE/TRIANGLE RELATIONSHIPS

6.2 Rates

Learning Objectives
1. Write rates as fractions.
2. Find unit rates.
3. Find the best buy based on cost per unit.

Key Terms

Use the vocabulary terms listed below to complete each statement in exercises 1−3.

    rate   **unit rate**   **cost per unit**

1. When the denominator of a rate is 1, it is called a _____.

2. The _____ is that rate that tells how much is paid for one item.

3. A _____ compares two measurements with different units.

Objective 1   Write rates as fractions.
Video Examples

Review these examples for Objective 1:
1. Write each rate as a fraction in lowest terms.

    a. 7 gallons for $56

    Write the units: gallons and dollars
    $\dfrac{7 \text{ gallons} \div 7}{56 \text{ dollars} \div 7} = \dfrac{1 \text{ gallon}}{8 \text{ dollars}}$

    b. 75 inches of growth in 15 weeks

    $\dfrac{75 \text{ inches} \div 15}{15 \text{ weeks} \div 15} = \dfrac{5 \text{ inches}}{1 \text{ week}}$

    c. 984 miles on 32 gallons of gas

    $\dfrac{984 \text{ miles} \div 8}{32 \text{ gallons} \div 8} = \dfrac{123 \text{ miles}}{4 \text{ gallons}}$

Now Try:
1. Write each rate as a fraction in lowest terms.

    a. $7 for 35 pages

    _____

    b. 300 strokes in 20 minutes

    _____

    c. 396 strawberries for 24 cakes

    _____

Name: Date:
Instructor: Section:

Objective 1 Practice Exercises

For extra help, see Example 1 on page 439 of your text.

Write each rate as a fraction in lowest terms.

1. 119 pills for 17 patients   1. _____

2. 28 dresses for 4 women   2. _____

3. 256 pages for 8 chapters   3. _____

Objective 2   Find unit rates.

Video Examples

Review these examples for Objective 2:
2. Find each unit rate.

   a. 478.5 miles on 14.5 gallons of gas

Write the rate as a fraction.
$$\frac{478.5 \text{ miles}}{14.5 \text{ gallons}}$$
Divide 478.5 by 14.5 to find the unit rate.
$$145\overline{)4785} = 33$$

$$\frac{478.5 \text{ miles} \div 14.5}{14.5 \text{ gallons} \div 14.5} = \frac{33 \text{ miles}}{1 \text{ gallon}}$$

The unit rate is 33 miles per gallon or 33 miles/gallon.

   c. $1240 in 8 days

$$\frac{1240 \text{ dollars}}{8 \text{ days}} \quad \text{Divide: } 8\overline{)1240} = 155$$

The unit rate is $155/day.

Now Try:
2. Find each unit rate.

   a. 294 miles on 10.5 gallons of gas

_____

   c. $580 in 4 days

_____

Name: Date:
Instructor: Section:

Objective 2 Practice Exercises

For extra help, see Example 2 on page 440 of your text.

Find each unit rate.

4. $3500 in 20 days   4. _____

5. $7875 for 35 pounds   5. _____

6. 189.88 miles on 9.4 gallons   6. _____

Objective 3   Find the best buy based on cost per unit.

Video Examples

Review these examples for Objective 3: | **Now Try:**
3. Determine the best price for peanut butter. For 18 ounces the price is $2.89, for 28 ounces the price is $3.99, and for 40 ounces, the price is $6.18.

For 18 ounces, the cost per ounce is
$$\frac{\$2.89}{18 \text{ ounces}} \approx \$0.16$$
For 28 ounces, the cost per ounce is
$$\frac{\$3.99}{28 \text{ ounces}} \approx \$0.14$$
For 40 ounces, the cost per ounce is
$$\frac{\$6.18}{40 \text{ ounces}} \approx \$0.15$$
The lowest cost per ounce is $0.14, so the 28-ounce container is the best buy.

3. Find the best buy:
   2 pints for $3.55,
   3 pints for $5.25, and
   5 pints for $8.50.

   _____

Name: Date:
Instructor: Section:

4a. Solve the application problem.

Brand AA laundry detergent costs $5.99 for 32 ounces. Brand ZZ laundry detergent costs $13.29 for 100 ounces. Which choice is the best buy?

To find Brand AA's unit cost, divide $5.99 by 32 ounces. Similarly, to find Brand ZZ's unit cost, divide $13.29 by 100 ounces.

Brand AA $\dfrac{\$5.99}{32 \text{ ounces}} \approx 0.187$ per ounce

Brand ZZ $\dfrac{\$13.29}{100 \text{ ounces}} = 0.1329$ per ounce

Brand ZZ has the lower cost per ounce and is the better buy.

4a. Solve the application problem.

An eight-pack of AA-size batteries costs $4.99. A twenty-pack of AA-size batteries costs $12.99. Which battery pack is the best buy?

_____

Objective 3 Practice Exercises

For extra help, see Examples 3–4 on pages 440–442 of your text.

Find the best buy (based on cost per unit) for each item.

7. Peanut butter: 18 ounces for $1.77; 24 ounces for $2.08

7. _____

8. Batteries: 4 for $2.79; 10 for $4.19

8. _____

9. Soup: 3 cans for $1.75; 5 cans for $2.75; 8 cans for $4.55

9. _____

Name: Date:
Instructor: Section:

Chapter 6 RATIO, PROPORTION, AND LINE/ANGLE/TRIANGLE RELATIONSHIPS

6.3 Proportions

Learning Objectives
1. Write proportions.
2. Determine whether proportions are true or false.
3. Find the unknown number in a proportion.

Key Terms

Use the vocabulary terms listed below to complete each statement in exercises 1–3.

    cross products   **proportion**   **ratio**

1. A _____ shows that two ratios or rates are equivalent.

2. To see whether a proportion is true, determine if the _____ are equal.

3. A _____ is a comparison of two quantities with the same units.

Objective 1   Write proportions.

Video Examples

Review these examples for Objective 1:
1. Write each proportion.

   a. 7 m is to 13 m as 28 m is to 52 m

   $\dfrac{7 \, \cancel{m}}{13 \, \cancel{m}} = \dfrac{28 \, \cancel{m}}{52 \, \cancel{m}}$ so $\dfrac{7}{13} = \dfrac{28}{52}$

   b. $14 is to 8 gallons as $7 is to 4 gallons

   $\dfrac{\$14}{8 \text{ gallons}} = \dfrac{\$7}{4 \text{ gallons}}$

Now Try:
1. Write each proportion.

   a. 24 ft is to 17 ft as 72 ft is to 51 ft

   _____

   b. $10 is to 7 cans as $60 is to 42 cans

   _____

Objective 1 Practice Exercises

For extra help, see Example 1 on page 447 of your text.

Write each proportion.

1. 50 is to 8 as 75 is to 12.   1. _____

Name: Date:
Instructor: Section:

2.   36 is to 45 as 8 is to 10.   2. _____

3.   3 is to 33 as 12 is to 132.   3. _____

Objective 2   Determine whether proportions are true or false.

Video Examples

Review these examples for Objective 2:

2. Determine whether each proportion is true or false by writing both ratios in lowest terms.

  a. $\dfrac{7}{11} = \dfrac{16}{24}$

  Write each ratio in lowest terms.

  $\dfrac{7}{11}$ ← Already in lowest terms $\dfrac{16 \div 8}{24 \div 8} = \dfrac{2}{3}$ ← Lowest terms

  Because $\dfrac{7}{11}$ is not equivalent to $\dfrac{2}{3}$, the proportion is false.

  b. $\dfrac{9}{15} = \dfrac{21}{35}$

  Write each ratio in lowest terms.

  $\dfrac{9 \div 3}{15 \div 3} = \dfrac{3}{5}$ $\dfrac{21 \div 7}{35 \div 7} = \dfrac{3}{5}$

  Both ratios are equivalent to $\dfrac{3}{5}$, so the proportion is true.

3. Use cross products to see whether each proportion is true or false.

  a. $\dfrac{5}{8} = \dfrac{30}{48}$

  Multiply along one diagonal, then multiply along the other diagonal.

  $\dfrac{5}{8} = \dfrac{30}{48}$ $\nearrow 8 \cdot 30 = 240$
                                       $\searrow 5 \cdot 48 = 240$

  The cross products are equal, so the proportion is true.

Now Try:

2. Determine whether each proportion is true or false by writing both ratios in lowest terms.

  a. $\dfrac{36}{28} = \dfrac{24}{18}$

  _____

  b. $\dfrac{4}{12} = \dfrac{9}{27}$

  _____

3. Use cross products to see whether each proportion is true or false.

  a. $\dfrac{6}{17} = \dfrac{18}{51}$

  _____

Name: Date:
Instructor: Section:

b. $\dfrac{3\frac{1}{5}}{4\frac{2}{3}} = \dfrac{7}{10}$

$\dfrac{3\frac{1}{5}}{4\frac{2}{3}} = \dfrac{7}{10}$

$\nearrow 4\frac{2}{3} \cdot 7 = \dfrac{14}{3} \cdot \dfrac{7}{1} = \dfrac{98}{3} = 32\frac{2}{3}$

$\searrow 3\frac{1}{5} \cdot 10 = \dfrac{16}{\cancel{5}} \cdot \dfrac{\cancel{10}^{2}}{1} = \dfrac{32}{1} = 32$

The cross products are unequal, so the proportion is false.

b. $\dfrac{3.2}{5} = \dfrac{7}{10}$

_____

Objective 2 Practice Exercises

For extra help, see Examples 2–3 on pages 447–449 of your text.

Determine whether each proportion is true or false by writing the ratios in lowest terms. Show the simplified ratios and then write **true** *or* **false**.

4. $\dfrac{48}{36} = \dfrac{3}{4}$

4. _____

5. $\dfrac{30}{25} = \dfrac{6}{5}$

5. _____

Use cross products to determine whether the proportion is true or false. Show the cross products and then write **true** *or* **false**.

6. $\dfrac{4\frac{3}{5}}{9} = \dfrac{18\frac{2}{5}}{36}$

6. _____

Name: Date:
Instructor: Section:

Objective 3   Find the unknown number in a proportion.

Video Examples

Review these examples for Objective 3:

4. Find the unknown number in each proportion. Round answers to the nearest hundredth when necessary.

   a. $\dfrac{14}{x} = \dfrac{21}{18}$

   Recall that ratios can be rewritten in lowest terms. Write $\dfrac{21}{18}$ in lowest terms as $\dfrac{7}{6}$, which gives the proportion $\dfrac{14}{x} = \dfrac{7}{6}$.

   Step 1 Find the cross product.
   $$\dfrac{14}{x} = \dfrac{7}{6} \quad \nearrow x \cdot 7 \quad \searrow 14 \cdot 6$$

   Step 2 Show that cross products are equal.
   $x \cdot 7 = 14 \cdot 6$
   $x \cdot 7 = 84$

   Step 3 Divide both sides by 7.
   $$\dfrac{x \cdot \cancel{7}}{\cancel{7}} = \dfrac{84}{7}$$
   $x = 12$

   Step 4 Check in original proportion.
   $$\dfrac{14}{12} = \dfrac{21}{18} \quad \nearrow 12 \cdot 21 = 252 \quad \searrow 14 \cdot 18 = 252$$

   The cross products are equal, so 12 is the correct solution.

   b. $\dfrac{9}{13} = \dfrac{21}{x}$

   Step 1 Find the cross product.
   $$\dfrac{9}{13} = \dfrac{21}{x} \quad \nearrow 13 \cdot 21 = 273 \quad \searrow 9 \cdot x$$

   Step 2 Show that cross products are equal.
   $9 \cdot x = 273$

Now Try:

4. Find the unknown number in each proportion. Round answers to the nearest hundredth when necessary.

   a. $\dfrac{24}{x} = \dfrac{9}{12}$

   _____

   b. $\dfrac{2}{3} = \dfrac{x}{16}$

   _____

Name: Date:
Instructor: Section:

Step 3 Divide both sides by 9.

$$\frac{\cancel{9} \cdot x}{\cancel{9}} = \frac{273}{9}$$

$x = 30.33$ rounded to the nearest hundredth

Step 4 Check in original proportion.

$$\frac{9}{13} = \frac{21}{30.33} \quad \begin{matrix} \nearrow 13 \cdot 21 = 273 \\ \searrow 9 \cdot 30.33 = 272.97 \end{matrix}$$

The cross products are slightly different because of the rounded value of x. However, they are close enough to see that the problem was done correctly and that 30.33 is the approximate solution.

5a. Find the unknown number in the proportion.

$$\frac{3\frac{1}{4}}{8} = \frac{x}{12}$$

$$\frac{3\frac{1}{4}}{8} = \frac{x}{12} \quad \begin{matrix} \nearrow 8 \cdot x \\ \searrow 3\frac{1}{4} \cdot 12 \end{matrix}$$

Change $3\frac{1}{4}$ to an improper fraction and write in lowest terms.

$$3\frac{1}{4} \cdot 12 = \frac{13}{4} \cdot \frac{12}{1} = \frac{13}{\cancel{4}} \cdot \frac{\cancel{12}^3}{1} = \frac{39}{1} = 39$$

Show that the cross products are equal.
    $8 \cdot x = 39$

Divide both sides by 8.

$$\frac{\cancel{8} \cdot x}{\cancel{8}} = \frac{39}{8}$$

Write the solution as a mixed number in lowest terms.

$$x = \frac{39}{8} = 4\frac{7}{8}$$

The unknown number is $4\frac{7}{8}$.

5a. Find the unknown number in the proportion.

$$\frac{4\frac{1}{3}}{5} = \frac{x}{3}$$

_____

Name: Date:
Instructor: Section:

Check in original proportion.

$$\frac{3\frac{1}{4}}{8} = \frac{4\frac{7}{8}}{12}$$

$$8 \cdot 4\frac{7}{8} = \frac{\cancel{8}^1}{1} \cdot \frac{39}{\cancel{8}_1} = 39$$

$$3\frac{1}{4} \cdot 12 = \frac{13}{\cancel{4}_1} \cdot \frac{\cancel{12}^3}{1} = 39$$

The cross products are equal, so $4\frac{7}{8}$ is the correct solution.

Objective 3 Practice Exercises

For extra help, see Examples 4–5 on pages 450–453 of your text.

Find the unknown number in the proportion.

7. $\dfrac{9}{7} = \dfrac{x}{28}$ 7. _____

Find the unknown number in each proportion.

8. $\dfrac{2}{3\frac{1}{4}} = \dfrac{8}{x}$ 8. _____

9. $\dfrac{3}{x} = \dfrac{0.8}{5.6}$ 9. _____

Name:　　　　　　　　　　　　　　　Date:
Instructor:　　　　　　　　　　　　　Section:

Chapter 6 RATIO, PROPORTION, AND LINE/ANGLE/TRIANGLE RELATIONSHIPS

6.4　Problem Solving with Proportions

Learning Objectives
1　Use proportions to solve application problems.

Key Terms

Use the vocabulary terms listed below to complete each statement in exercises 1–2.

　　rate　　　　　ratio

1. A statement that compares a number of inches to a number of inches is a
   _____.

2. A statement that compares a number of gallons to a number of miles is a
   _____.

Objective 1　Use proportions to solve application problems.

Video Examples

Review these examples for Objective 1:

1. Alexis drove 343 miles on 7.5 gallons of gas. How far can she travel on a full tank of 12 gallons of gas?

   Step 1 Read the problem. The problem asks for the number of miles the car can travel on 12 gallons of gas.

   Step 2 Work out a plan. Decide what is being compared and write a proportion using the two rates.

   $$\frac{343 \text{ miles}}{7.5 \text{ gallons}} = \frac{x \text{ miles}}{12 \text{ gallons}}$$

   Step 3 Estimate a reasonable answer. To estimate the answer, notice that 12 is a little more than 1.5 times 7.5. So use $343 \cdot 1.5 = 514.5$ miles, as an estimate.

   Step 4 Solve the problem. Ignore the units while solving for x.

   $$\frac{343 \text{ miles}}{7.5 \text{ gallons}} = \frac{x \text{ miles}}{12 \text{ gallons}}$$
   $$(7.5)(x) = (343)(12)$$
   $$(7.5)(x) = 4116$$

Now Try:

1. Aiden spends 23 hours painting 4 apartments. How long will it take him to paint the other 16 apartments?

   _____

Copyright © 2018 Pearson Education, Inc.　　　　　　197

Name: Date:
Instructor: Section:

$$\frac{\cancel{(7.5)}(x)}{\cancel{7.5}} = \frac{4116}{7.5}$$

$$x = 548.8 \quad \text{Round to 549.}$$

Step 5 State the answer. Rounded to the nearest mile, the car can travel about 549 miles on a full tank of gas.

Step 6 Check your work. The answer 549 miles, is a little more than the estimate of 514.5 miles, so it is reasonable.

2. There are 24 women in a college class of 39. At that rate, how many of the college's 10,400 students are women?

   Step 1 Read the problem. The problem asks how many of the 10,400 students are women.

   Step 2 Work out a plan. Decide what is being compared and write a proportion using the two rates.

   $$\frac{24 \text{ women}}{39 \text{ students}} = \frac{x \text{ women}}{10{,}400 \text{ students}}$$

   Step 3 Estimate a reasonable answer. To estimate the answer, notice that 24 is a little more than half 39. Half of 10,400 is $10{,}400 \div 2 = 5200$, so our estimate is more than 5200 students.

   Step 4 Solve the problem. Ignore the units while solving for *x*.

   $$\frac{24 \text{ women}}{39 \text{ students}} = \frac{x \text{ women}}{10{,}400 \text{ students}}$$
   $$39 \cdot x = 24 \cdot 10{,}400$$
   $$39 \cdot x = 249{,}600$$
   $$\frac{\cancel{39} \cdot x}{\cancel{39}} = \frac{249{,}600}{39}$$
   $$x = 6400 \quad \text{No need to round.}$$

   Step 5 State the answer. There are 6400 woman at the college.

   Step 6 Check your work. The answer 6400 women, is a little more than the estimate of 5200 women, so it is reasonable.

2. A survey showed that 4 out of 5 smokers have tried to quit smoking. At this rate, how many people in a group of 540 have tired to quit smoking?

   _____

Name: Date:
Instructor: Section:

Objective 1 Practice Exercises

For extra help, see Examples 1–2 on pages 460–462 of your text.

Set up and solve a proportion for each problem.

1. If 22 hats cost $198, find the cost of 12 hats.

    1. _____

2. If 150 square yards of carpet cost $3142.50, find the cost of 210 square yards of the carpet.

    2. _____

3. A biologist tags 50 deer and releases them in a wildlife preserve area. Over the course of a two-week period, she observes 80 deer, of which 12 are tagged. What is the estimate for the population of deer in this particular area?

    3. _____

Name: Date:
Instructor: Section:

Chapter 6 RATIO, PROPORTION, AND LINE/ANGLE/TRIANGLE RELATIONSHIPS

6.5 Geometry: Lines and Angles

Learning Objectives	
1	Identify and name lines, line segments, and rays.
2	Identify parallel and intersecting lines.
3	Identify and name angles.
4	Classify angles as right, acute, straight, or obtuse.
5	Identify perpendicular lines.
6	Identify complementary angles and supplementary angles, and find the measure of a complement or supplement of a given angle.
7	Identify congruent angles and vertical angles, and use this knowledge to find the measures of angles.
8	Identify corresponding angles and alternate interior angles, and use this knowledge to find the measures of angles.

Key Terms

Use the vocabulary terms listed below to complete each statement in exercises 1–19.

point	**line**	**line segment**	**ray**
parallel lines	**intersecting lines**	**angle**	
degrees	**straight angle**	**right angle**	
acute angle	**obtuse angle**	**perpendicular lines**	
complementary angles	**supplementary angles**		
congruent angles	**vertical angles**		
corresponding angles	**alternate interior angles**		

1. A _____ is a part of a line that has one endpoint and which extends infinitely in one direction.

2. Two lines that intersect to form a right angle are _____.

3. An angle whose measure is between 90° and 180° is an _____.

4. A _____ is a location in space.

5. Two rays with a common endpoint form an _____.

6. A set of points that form a straight path that extends infinitely in both directions is called a _____.

7. An angle that measures less than 90° is called an _____.

Name: Date:
Instructor: Section:

8. Angles are measured using _____.

9. Two lines in the same plane that never intersect are _____.

10. A part of a line with two endpoints is a _____.

11. Two lines that cross at one point are _____.

12. An angle whose measure is exactly 90º is a _____.

13. The nonadjacent angles formed by two intersecting lines are called _____.

14. Angles whose measures are equal are called _____.

15. Two angles whose measures sum to 180º are _____.

16. Two angles whose measures sum to 90º are _____.

17. When two parallel lines are cut by a transversal, the angles between the parallel lines on opposite sides of the transversal are called _____.

18. When two parallel lines are cut by a transversal, the angles in the same relative position with regard to the parallel lines and the transversal are called _____.

19. An angle whose measure is exactly 180º is a _____.

Objective 1 Identify and name lines, line segments, and rays.

Video Examples

Review these examples for Objective 1:
1. Identify each figure below as a line, line segment, or ray, and name it using the appropriate symbol.

   a.

   This figure has two endpoints, so it is a line segment named $\overline{GH}$ or $\overline{HG}$.

Now Try:
1. Identify each figure below as a line, line segment, or ray, and name it using the appropriate symbol.

   a.
   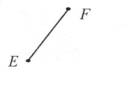

   _____

Copyright © 2018 Pearson Education, Inc.

Name: Date:
Instructor: Section:

b.

This figure starts at point *B* and goes on forever in one direction, so it is a ray named $\overrightarrow{BA}$.

c.

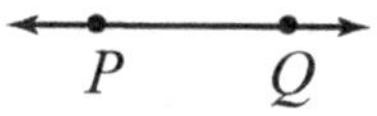

$\overset{\longleftrightarrow}{P\ Q}$

This figure goes on forever in both directions, so it is a line named $\overleftrightarrow{PQ}$ or $\overleftrightarrow{QP}$.

b.

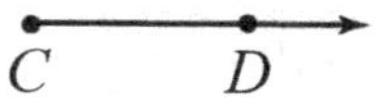

$\overset{\longrightarrow}{C\ D}$

c.

$\overset{\longleftrightarrow}{R\ S}$

Objective 1 Practice Exercises

For extra help, see Example 1 on page 467 of your text.

Identify each figure as a line, line segment, or ray, and name it.

1.

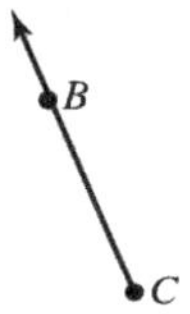

1. _____

2.
 (V above W with arrows both directions)

2.

3.

1. _____

2. _____

3. _____

Name: Date:
Instructor: Section:

Objective 2 Identify parallel and intersecting lines.

Video Examples

Review these examples for Objective 2:

2. Label each pair of the lines as appearing to be parallel or as intersecting.

   a.

   The lines in this figure cross, so they are intersecting lines.

   b.

   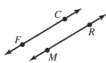

   The lines in this figure do not intersect; they appear to be parallel.

Now Try:

2. Label each pair of the lines as appearing to be parallel or as intersecting.

   a.

   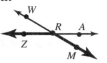

   _____

   b.

   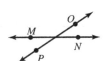

   _____

Objective 2 Practice Exercises

For extra help, see Example 2 on page 468 of your text.

*Label each pair of lines as appearing to be **parallel** or **intersecting**.*

4.

4. _____

5.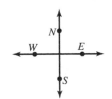

5. _____

6.

6. _____

Copyright © 2018 Pearson Education, Inc.   203

Name: Date:
Instructor: Section:

Objective 3 Identify and name angles.

Video Examples

Review this example for Objective 3:

3. Name the highlighted angle in two different ways.

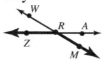

The angle can be named $\angle ZRM$ or $\angle MRZ$. It cannot be named $\angle R$, using the vertex alone, because four different angles have R as their vertex.

Now Try:

3. Name the highlighted angle in two different ways.

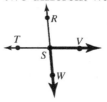

_____

Objective 3 Practice Exercises

For extra help, see Example 3 on page 469 of your text.

Name each angle drawn with darker rays by using the three-letter form of identification.

7.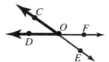

7. _____

8.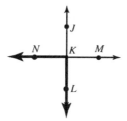

8. _____

9.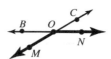

9. _____

Name: Date:
Instructor: Section:

Objective 4 Classify angles as right, acute, straight, or obtuse.

Video Examples

Review these examples for Objective 4:

4. Label each angle as acute, right, obtuse, or straight.

   a.

   This figure shows a straight angle (exactly 180°).

   b.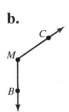

   This figure shows an obtuse angle (more than 90° but less than 180°).

   c.

   This figure shows an acute angle (less than 90°).

   d.

   This figure shows a right angle (exactly 90° and identified by a small square at the vertex).

Now Try:

4. Label each angle as acute, right, obtuse, or straight.

   a.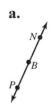

   _____

   b.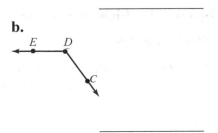

   _____

   c.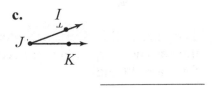

   _____

   d.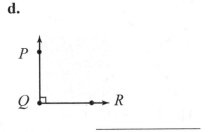

   _____

Objective 4 Practice Exercises

For extra help, see Example 4 on page 470 of your text.

Label each angle as **acute**, **right**, **obtuse**, *or* **straight**.

10.

10. _____

Name: Date:
Instructor: Section:

11. 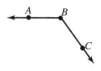 11. _____

12.   12. _____

Objective 5 Identify perpendicular lines.
Video Examples

Review these examples for Objective 5:
5. Which pairs of lines are perpendicular?

   a.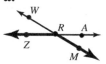

   The lines in this figure are intersecting lines, but they are not perpendicular because they do not form a right angle.

   b.

   The lines in this figure are perpendicular to each other because they intersect at right angles.

Now Try:
5. Which pairs of lines are perpendicular?

   a.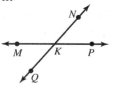

   _____

   b.

   _____

Objective 5 Practice Exercises

For extra help, see Example 5 on page 471 of your text.

Label each pair of lines as appearing to be **parallel**, **perpendicular**, *or* **intersecting**.

13. 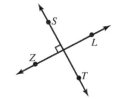 13. _____

206 Copyright © 2018 Pearson Education, Inc.

Name: Date:
Instructor: Section:

14.

14. _____

15.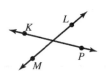

15. _____

Objective 6 Identify complementary angles and supplementary angles, and find the measure of a complement or supplement of a given angle.

Video Examples

Review these examples for Objective 6:

6. Identify each pair of complementary angles.

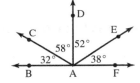

$\angle BAC\,(32°)$ and $\angle CAD\,(58°)$ are complementary angles because
$32° + 58° = 90°$

$\angle DAE\,(52°)$ and $\angle EAF\,(38°)$ are complementary angles because
$52° + 38° = 90°$

Now Try:

6. Identify each pair of complementary angles.

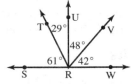

_____

8. In the figures below, $\angle ABC$ measures 150°, $\angle WXY$ measures 30°, $\angle MKQ$ measures 30°, and $\angle QKP$ measures 150°. Identify each pair of supplementary angles.

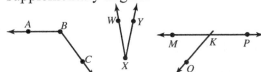

$\angle ABC$ and $\angle WXY$, because $150° + 30° = 180°$
$\angle MKQ$ and $\angle QKP$, because $30° + 150° = 180°$
$\angle ABC$ and $\angle MKQ$, because $150° + 30° = 180°$
$\angle WXY$ and $\angle QKP$, because $30° + 150° = 180°$

8. In the figures below, $\angle OQP$ measures 35°, $\angle PQN$ measures 145°, $\angle RST$ measures 35°, and $\angle BMC$ measures 145°. Identify each pair of supplementary angles.

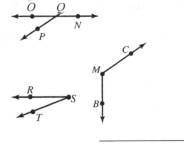

_____

Name: Date:
Instructor: Section:

Objective 6 Practice Exercises

For extra help, see Examples 6–9 on pages 471–472 of your text.

Find the complement of the angle.

16. 43°

16. _____

Find the supplement of the angle.

17. 16°

17. _____

Identify each pair of supplementary angles.

18.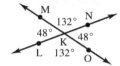

18. _____

Objective 7 Identify congruent angles and vertical angles, and use this knowledge to find the measures of angles.

Video Examples

Review this example for Objective 7:
10. Identify the angles that are congruent.

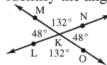

Congruent angles measure the same number of degrees.
∠LKM ≅ ∠OKN and ∠LKO ≅ ∠MKN.

Now Try:
10. Identify the angles that are congruent.

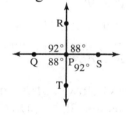

_____

Objective 7 Practice Exercises

For extra help, see Examples 10–12 on pages 473–474 of your text.

In the figure below, identify the angles that are congruent.

19.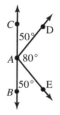

19. _____

Name: Date:
Instructor: Section:

In the figure below, identify all the vertical angles.

20.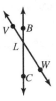

20. _____

In the figure below, ∠ABE measures 73° and ∠FEB measures 107°. Find the measures of the indicated angles.

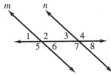

21.   ∠CBG and ∠DEH

21. _____

Objective 8   Identify corresponding angles and alternate interior angles, and use this knowledge to find the measures of angles.

Video Examples

Review these examples for Objective 8:
13. In each figure, line *m* is parallel to line *n*.

a. Identify all pairs of corresponding angles.

There are four pair of corresponding angles:
∠1 and ∠3   ∠2 and ∠4
∠5 and ∠7   ∠6 and ∠8

b. Identify all pairs of alternate interior angles.

There are two pair of alternate interior angles:
∠2 and ∠7
∠6 and ∠3

Now Try:
13. In each figure, line *m* is parallel to line *n*. Identify all pairs of corresponding angles and all pairs of alternate interior angles.

a. Identify all pairs of corresponding angles.

_____

_____

b. Identify all pairs of alternate interior angles.

_____

_____

Name: Date:
Instructor: Section:

14. In the figure below, line *m* is parallel to line *n* and the measure of ∠3 is 44°. Find the measure of the other angles.

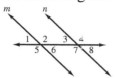

We know that ∠3 is 44°.
   ∠3 ≅ ∠1 (corresponding angles), so the measure of ∠1 is also 44°.
   ∠3 ≅ ∠6 (alternate interior angles), so the measure of ∠6 is also 44°.
   ∠6 ≅ ∠8 (corresponding angles), so the measure of ∠8 is also 44°.
Notice that the exterior sides of ∠3 and ∠4 form a straight line. Therefore, ∠3 and ∠4 are supplementary angles. If ∠3 is 44°, then ∠4 must be 180° – 44° = 136°.
   ∠4 ≅ ∠2 (corresponding angles), so the measure of ∠4 is also 136°.
   ∠7 ≅ ∠2 (alternate interior angles), so the measure of ∠2 is also 136°.
   ∠7 ≅ ∠5 (corresponding angles), so the measure of ∠5 is also 136°.

14. In the figure below, line *m* is parallel to line *n* and the measure of ∠6 is 125°. Find the measure of the other angles.

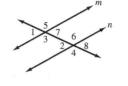

_____

_____

_____

_____

Objective 8 Practice Exercises

For extra help, see Examples 13–14 on pages 475–476 of your text.

In each figure, line m is parallel to line n. List the corresponding angles and the alternate interior angles. Then find the measure of each angle.

22. ∠4 measures 100°.

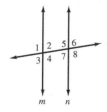

22.
Corresponding angles:

_____

Alternate interior angles:

_____

Angle measures:

_____

23. ∠7 measures 37°.

23.
Corresponding angles:

_____

Alternate interior angles:

_____

Angle measures:

_____

Name: Date:
Instructor: Section:

Chapter 6 RATIO, PROPORTION, AND LINE/ANGLE/TRIANGLE RELATIONSHIPS

6.6 Geometry Applications: Congruent and Similar Triangles

Learning Objectives
1. Identify corresponding parts of congruent triangles.
2. Prove that triangles are congruent using ASA, SSS, or SAS.
3. Identify corresponding parts of similar triangles.
4. Find the unknown lengths of sides in similar triangles.
5. Solve application problems involving similar triangles.

Key Terms

Use the vocabulary terms listed below to complete each statement in exercises 1–4.

    congruent figures   similar figures

    congruent triangles   similar triangles

1. Triangles with the same shape and size are _____.

2. _____ are identical both in shape and in size.

3. Triangles with the same shape but not necessarily the same size are _____.

4. _____ have the same shape but are different sizes.

Objective 1   Identify corresponding parts of congruent triangles.

Video Examples

Review this example for Objective 1:	Now Try:
1. Identify corresponding angles and sides in these congruent triangles. 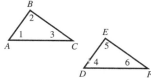 The corresponding parts are: ∠1 and ∠4 $\overline{AB}$ and $\overline{DE}$ ∠2 and ∠5 $\overline{BC}$ and $\overline{EF}$ ∠3 and ∠6 $\overline{AC}$ and $\overline{DF}$	1. Identify corresponding angles and sides in these congruent triangles. 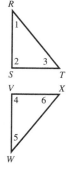 _____

Name: Date:
Instructor: Section:

Objective 1 Practice Exercises

For extra help, see Example 1 on pages 481–482 of your text.

Each pair of triangles is congruent. List the corresponding angles and the corresponding sides.

1.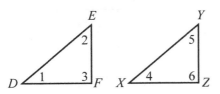

1. _____

2.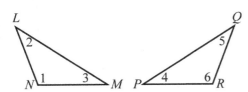

2. _____

Objective 2 Prove that triangles are congruent using ASA, SSS, or SAS.

Video Examples

Review this example for Objective 2:

2a. Explain which method can be used to prove that the pair of triangles is congruent. Choose from ASA, SSS, and SAS.

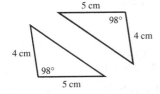

On both triangles, two corresponding sides and the angle between them measure the same, so Side-Angle-Side (SAS) method can be used to prove that the triangles are congruent.

Now Try:

2a. Explain which method can be used to prove that the pair of triangles is congruent. Choose from ASA, SSS, and SAS.

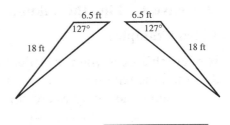

_____

Copyright © 2018 Pearson Education, Inc.   213

Name: Date:
Instructor: Section:

Objective 2 Practice Exercises

For extra help, see Example 2 on page 483 of your text.

Determine which of these methods can be used to prove that each pair of triangles is congruent: Angle-Side-Angle (ASA), Side-Side-Side (SSS), or Side-Angle-Side (SAS).

3. 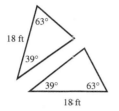   3. _____

4. 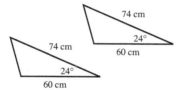   4. _____

5. 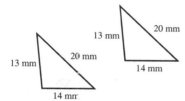   5. _____

Objective 3 Identify the corresponding parts of similar triangles.

For extra help, see page 483 of your text.

Objective 4 Find the unknown lengths of sides in similar triangles.

Video Examples

Review this example for Objective 4:
3. Find the length of *b* in the smaller triangle. Assume the triangles are similar.

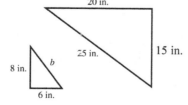

The length you want to find in the smaller triangle is side *b*, and it corresponds to 25 in. in the larger triangle. The smaller triangle is turned

Now Try:
3. Find the length of *x* in the smaller triangle. Assume the triangles are similar.

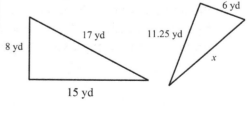

_____

214   Copyright © 2018 Pearson Education, Inc.

"upside down" compared to the larger triangle, so be careful when identifying corresponding sides. Then notice that 6 in. in the smaller triangle corresponds to 15 in. in the larger triangle, and you know both of their lengths. Because the ratios of the lengths of corresponding sides are equal, you can set up a proportion.

$$\frac{b}{25} = \frac{6}{15}$$

$$\frac{b}{25} = \frac{2}{5} \quad \text{Write } \frac{6}{15} \text{ in lowest terms as } \frac{2}{5}.$$

Find the cross products.

$$b \cdot 5 = 25 \cdot 2$$

$$\frac{b \cdot \cancel{5}}{\cancel{5}} = \frac{50}{5}$$

$$b = 10$$

Side b has length of 10 in.

Objective 4 Practice Exercises

For extra help, see Examples 3–4 on pages 484–485 of your text.

Find the unknown lengths in each pair of similar triangles.

6.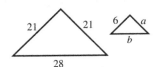

6. _____

7.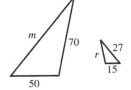

7. _____

Name: Date:
Instructor: Section:

Find the perimeter of each triangle. Assume the triangles are similar.

8. 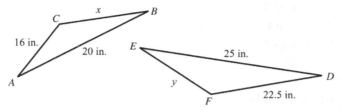   8. ABC _____

                                                                  EDF _____

Objective 5   Solve application problems involving similar triangles.

Video Examples

Review this example for Objective 5:

5. The height of the house shown here can be found by using similar triangles and proportion. Find the height of the house by writing a proportion and solving it.

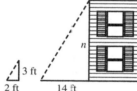

The triangles shown are similar, so write a proportion to find n.

height of larger triangle → $\dfrac{n}{3} = \dfrac{14}{2}$ ← length of larger triangle
height of smaller triangle   ← length of smaller triangle

Find the cross products and show that they are equal.

$$n \cdot 2 = 3 \cdot 14$$
$$n \cdot 2 = 42$$
$$\dfrac{n \cdot \cancel{2}}{\cancel{2}} = \dfrac{42}{2}$$
$$n = 21$$

The height of the house is 21 ft.

Now Try:

5. A flagpole casts a shadow 77 feet long at the same time that a pole 15 feet tall casts a shadow 55 ft long. Find the height of the flagpole.

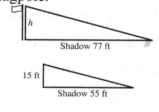

_____

Name: Date:
Instructor: Section:

Objective 5 Practice Exercises

For extra help, see Example 5 on page 486 of your text.

Solve each application problem.

9. A sailor on the USS Ramapo saw one of the highest waves ever recorded. He used the height of the ship's mast, the length of the deck and similar triangles to find the height of the wave. Using the information in the figure, write a proportion and then find the height of the wave.

9.

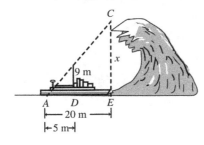

10. A fire lookout tower provides an excellent view of the surrounding countryside. The height of the tower can be found by lining up the top of the tower with the top of a 3-meter stick. Use similar triangles to find the height of the tower.

10.

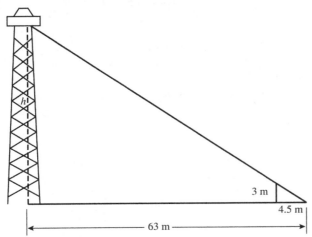

Copyright © 2018 Pearson Education, Inc.   217

11. A 30 m ladder touches the side of a building at a height of 25 m. At what height would a 12-m ladder touch the building if it makes the same angle with the ground?

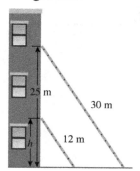

11. _____

Name: Date:
Instructor: Section:

Chapter 7 PERCENT

7.1 The Basics of Percent

Learning Objectives
1. Learn the meaning of percent.
2. Write percents as decimals.
3. Write decimals as percents.
4. Write percents as fractions.
5. Write fractions as percents.
6. Use 100% and 50%.

Key Terms

Use the vocabulary terms listed below to complete each statement in exercises 1–3.

    percent   ratio   decimals

1. To compare two quantities that have the same type of units, use a _____.

2. _____ means per one hundred.

3. _____ represent parts of a whole.

Objective 1   Learn the meaning of percent.

Video Examples

Review these examples for Objective 1:	Now Try:
1. Write a percent to describe each situation. **a.** You leave a $25 tip when the restaurant bill was $100. What percent is the tip? The tip is $25 per $100, or $\frac{25}{100}$. The percent is 25%. **b.** You pay $8 tax on a $100 catering order. What is the tax rate? The tax is $8 per $100, or $\frac{8}{100}$. The percent is 8%.	1. Write a percent to describe each situation. **a.** You give $10 to a charity per $100. What is the percent given to charity? _____ **b.** The tax on a hotel room of $100 is $18. What is the tax rate? _____

Name: Date:
Instructor: Section:

Objective 1 Practice Exercises

For extra help, see Example 1 on page 510 of your text.

Write as a percent.

1. 68 people out of 100 drive small cars.   1. _____

2. The tax is $9 per $100.   2. _____

3. The cost for labor was $45 for every $100 spent to manufacture an item.   3. _____

Objective 2   Write percents as decimals.

Video Examples

Review these examples for Objective 2:

2. Write each percent as a decimal.

   a. 39%

   39% = 39 ÷ 100 = 0.39

   b. 7%

   7% = 7 ÷ 100 = 0.07

   c. 89.9%

   89.9% = 89.9 ÷ 100 = 0.899

   e. 169%

   169% = 169 ÷ 100 = 1.69

3. Write each percent as a decimal by dropping the percent symbol and moving the decimal point two places to the left.

   a. 56%

   Drop the percent sign and move the decimal point two places to the left.
   56% = 56.% = 0.56

   d. 0.2%

   Two zeros are attached so the decimal point can be moved two places to the left.
   0.2% = 0.002

Now Try:

2. Write each percent as a decimal.

   a. 23%

   _____

   b. 8%

   _____

   c. 15.3%

   _____

   e. 302%

   _____

3. Write each percent as a decimal by dropping the percent symbol and moving the decimal point two places to the left.
   a. 48%

   _____

   d. 0.8%

   _____

Name: Date:
Instructor: Section:

Objective 2 Practice Exercises

For extra help, see Examples 2–3 on pages 511–512 of your text.

Write each percent as a decimal.

4. 42%   4. _____

5. 310%   5. _____

6. 18.9%   6. _____

Objective 3   Write decimals as percents.

Video Examples

Review these examples for Objective 3:

4. Write each decimal as a percent by moving the decimal point two places to the right.

   a. 0.31

   Decimal point is moved two places to the right and percent symbol is attached.
   0.31 = 31%

   e. 4.7

   0 is attached so the decimal point can be moved two places to the right.
   4.7 = 4.70 = 470%

   b. 0.904

   0.904 = 90.4%

Now Try:

4. Write each decimal as a percent by moving the decimal point two places to the right.

   a. 0.43

   _____

   e. 2.9

   _____

   b. 0.751

   _____

Objective 3 Practice Exercises

For extra help, see Example 4 on pages 512–513 of your text.

Write each decimal as a percent.

7. 0.2   7. _____

8. 0.564   8. _____

9. 4.93   9. _____

Name: Date:
Instructor: Section:

Objective 4 Write percents as fractions.

Video Examples

Review these examples for Objective 4:

6a. Write the percent as a fraction in lowest terms.

62.5%

Write 62.5 over 100.

$$62.5\% = \frac{62.5}{100}$$

To get a whole number in the numerator, multiply the numerator and denominator by 10.

$$\frac{62.5}{100} = \frac{62.5(10)}{100(10)} = \frac{625}{1000}$$

Now write the fraction in lowest terms.

$$\frac{625}{1000} = \frac{625 \div 125}{1000 \div 125} = \frac{5}{8}$$

5c. Write the percent as a fraction or mixed number in lowest terms.

350%

$$350\% = \frac{350}{100} = \frac{350 \div 50}{100 \div 50} = \frac{7}{2} = 3\frac{1}{2}$$

Now Try:

6a. Write the percent as a fraction in lowest terms.
28.5%

_____

5c. Write the percent as a fraction or mixed number in lowest terms.
175%

_____

Objective 4 Practice Exercises

For extra help, see Examples 5–6 on pages 513–515 of your text.

Write each percent as a fraction or mixed number in lowest terms.

10. 140%

11. $18\frac{1}{3}\%$

12. 55.6%

10. _____

11. _____

12. _____

Name: Date:
Instructor: Section:

Objective 5 Write fractions as percents.

Video Examples

Review this example for Objective 5:
7. Write the fraction as a percent. Round to the nearest tenth as necessary.

    b. $\frac{5}{16}$

    $\frac{5}{16} = \left(\frac{5}{16}\right)(100\%) = \left(\frac{5}{16}\right)\left(\frac{100}{1}\%\right)$
    $= \left(\frac{5}{4 \cdot 4}\right)\left(\frac{4 \cdot 25}{1}\%\right)$
    $= \frac{125}{4}\% = 31\frac{1}{4}\%$

Now Try:
7. Write the fraction as a percent. Round to the nearest tenth as necessary.

    b. $\frac{15}{16}$

    _____

Objective 5 Practice Exercises

For extra help, see Example 7 on pages 515–517 of your text.

Write each fraction or mixed number as a percent. If you're using a calculator, first work each one by hand. Then use your calculator and round to the nearest tenth of a percent, if necessary.

13. $\frac{47}{50}$

13. _____

14. $\frac{11}{40}$

14. _____

15. $\frac{64}{75}$

15. _____

Copyright © 2018 Pearson Education, Inc.

Name: Date:
Instructor: Section:

Objective 6 Use 100% and 50%.

Video Examples

Review these examples for Objective 6:
8a. Fill in the blanks.

100% of $69 is _____ .

100% is all of the money. So 100% of $69 is $69.

9a. Fill in the blanks.

50% of $68 is _____ .

50% is half of the money. So 50% of $68 is $34.

Now Try:
8a. Fill in the blanks.

100% of $6.15 is _____ .

_____

9a. Fill in the blanks.

50% of $7.30 is _____ .

_____

Objective 6 Practice Exercises

For extra help, see Examples 8–9 on page 517 of your text.

Fill in the blanks. Remember that 100% is all of something and 50% is half of it.

16. 50% of 260 miles is _____ . 16. _____

17. 100% of $520 is _____ . 17. _____

18. 50% of 40 acres is _____ . 18. _____

Name: Date:
Instructor: Section:

Chapter 7 PERCENT

7.2 The Percent Proportion

Learning Objectives
1. Identify the percent, whole, and part.
2. Solve percent problems using the percent proportion.

Key Terms

Use the vocabulary terms listed below to complete each statement in exercises 1–3.

    percent proportion   **whole**   **part**

1. The _____ in a percent problem is the entire quantity.

2. The _____ in a percent problem is the portion being compared with the whole.

3. Part is to whole as percent is to 100 is called the _____.

Objective 1   Identify the percent, whole, and part.

Video Examples

Review these examples for Objective 1:

1. Find the percent in the following.

    a. 43% of 1100 people exercise daily.

    The percent is 43.

    b. $255 is 29 percent of what number?

    The percent is 29, because 29 appears with the word percent.

    c. What percent of 6500 tons is 275 tons?

    The word percent has no number with it, so the percent is the unknown part of the problem.

Now Try:

1. Find the percent in the following.

    a. 950 tons of trash is 59% of what number of tons?

    _____

    b. Find the amount of sales tax by multiplying $650 and 7.75 percent.

    _____

    c. If 2650 children attend preschool, what percent of 3280 children attend preschool?

    _____

Name: Date:
Instructor: Section:

2. Identify the whole in the following.

   a. 43% of 1100 people exercise daily.
   The whole is 1100.

   b. $255 is 29 percent of what number?
   The whole is the unknown part of the problem.

   c. What percent of 6500 tons is 275 tons?
   The whole is 6500.

3. Identify the part in the following.

   a. 43% of 1100 people exercise daily.
   The part is the unknown part of the problem.

   b. $255 is 29 percent of what number?
   The part is $255.

   c. What percent of 6500 tons is 275 tons?
   The part is 275.

2. Identify the whole in the following.
   a. Find the amount of sales tax by multiplying $650 and 7.75 percent.

   _____

   b. 950 tons of trash is 59% of what number of tons?

   _____

   c. If 2650 children attend preschool, what percent of 3280 children attend preschool?

   _____

3. Identify the part in the following.
   a. Find the amount of sales tax by multiplying $650 and 7.75 percent.

   _____

   b. 950 tons of trash is 59% of what number of tons?

   _____

   c. If 2650 children attend preschool, what percent of 3280 children attend preschool?

   _____

Objective 1 Practice Exercises

For extra help, see Examples 1–3 on pages 525–526 of your text.

In each of the following, identify the percent, the whole, and the part. Do not try to solve for any unknowns.

1. $50 is 250% of what number?

1. percent _____

   whole _____

   part _____

Name: Date:
Instructor: Section:

2. 336 students is what percent of 840 students?

2. percent _____

   whole _____

   part _____

3. The state sales tax is $7\frac{3}{4}$ percent of the $895 price.

3. percent _____

   whole _____

   part _____

Objective 2   Solve percent problems using the percent proportion.

Video Examples

Review these examples for Objective 2:

4. Find 24% of $650.

   Here the percent is 24 and the whole is $650. Now find the part. Let x represent the unknown part.
   $$\frac{x}{650} = \frac{24}{100} \text{ or } \frac{x}{650} = \frac{6}{25}$$
   Find the cross products in the proportion and show that they are equal.
   $$x \cdot 25 = 650 \cdot 6$$
   $$x \cdot 25 = 3900$$
   $$\frac{x \cdot 25}{25} = \frac{3900}{25}$$
   $$x = 156$$
   24% of $650 is $156.

Now Try:

4. Find 38% of 6500.

   _____

5. Use the percent proportion to answer this question.
   9 pounds is what percent of 450 pounds?

   The percent proportion is $\frac{\text{percent}}{100} = \frac{\text{part}}{\text{whole}}$. Set up the proportion using p as the variable representing the unknown percent. Then find the cross products.

5. Use the percent proportion to answer this question. $28 is what percent of $700?

   _____

Copyright © 2018 Pearson Education, Inc.

Name: Date:
Instructor: Section:

$$\frac{p}{100} = \frac{9}{450}$$
$$p \cdot 450 = 900$$
$$\frac{p \cdot \cancel{450}}{\cancel{450}} = \frac{900}{450}$$
$$p = 2$$

The percent is 2%. So 9 pounds is 2% of 450 pounds.

6. Use the percent proportion to answer this question.

    150 students is 75% of how many students?

The percent proportion is $\frac{\text{percent}}{100} = \frac{\text{part}}{\text{whole}}$.

Here the percent is 75%, the part is 150, and the whole is unknown.

$$\frac{75}{100} = \frac{150}{n}$$
$$75 \cdot n = 15,000$$
$$\frac{\cancel{75} \cdot n}{\cancel{75}} = \frac{15,000}{75}$$
$$n = 200$$

The whole is 200 students. So 150 students is 75% of 200 students.

7b. Use the percent proportion to answer the question.

What percent of $40 is $46?

Here, the percent is unknown, the part is $46, and the whole is $40. Use the percent proportion.

$$\frac{p}{100} = \frac{46}{40}$$
$$p \cdot 40 = 4600$$
$$\frac{p \cdot \cancel{40}}{\cancel{40}} = \frac{4600}{40}$$
$$p = 115$$

The percent is 115%.

6. Use the percent proportion to answer this question. 45 cars is 30% of how many cars?

    _____

7b. Use the percent proportion to answer the question.

185% of what amount is $173.90?

    _____

Name: Date:
Instructor: Section:

Objective 2 Practice Exercises

For extra help, see Examples 4–7 on pages 526–529 of your text.

Use the percent proportion to answer these questions. If necessary, round money answers to the nearest cent and percent answers to the nearest tenth of a percent.

4. 165% of what number of feet is 12.7 feet? Round the answer to the nearest tenth.

   4. _____

5. What percent of 78 bicyclists is 110 bicyclists?

   5. _____

6. What number is 43% of 2200?

   6. _____

Name: Date:
Instructor: Section:

Chapter 7 PERCENT

7.3 The Percent Equation

Learning Objectives
1. Estimate answers to percent problems involving 25%.
2. Find 10% and 1% of a number by moving the decimal point.
3. Solve basic percent problems using the percent equation.

Key Terms

Use the vocabulary terms listed below to complete each statement in exercises 1–2.

    percent equation   percent

1. A number written with a _____ sign means "divided by 100".

2. The _____ is part = percent · whole.

Objective 1   Estimate answers to percent problems involving 25%.

Video Examples

Review these examples for Objective 1:
1. Estimate the answer to each question.

   a. What is 25% of $423?

   Use front end rounding to round $423 to $400. Then divide $400 by 4. The estimate is $100.

   b. Find 25% of 82.7 miles.

   Use front end rounding to round 82.7 miles to 80 miles. Then divide 80 by 4. The estimate is 20 miles.

   c. 25% of 89 days is how long?

   Round 89 days to 90 days using front end rounding. Then divide 90 by 4 to get 22.5 days.

Now Try:
1. Estimate the answer to each question.
   a. Find 25% of $205.

   _____

   b. What is 25% of 59 hours?

   _____

   c. 25% of 97 pounds is how many pounds?

   _____

Objective 1 Practice Exercises

For extra help, see Example 1 on page 532 of your text.

Estimate the answer to each question.

1. Find 25% of 97 minutes.   1. _____

Name: Date:
Instructor: Section:

2. 25% of 43 inches is how many inches? 2. _____

3. What is 25% of $7823? 3. _____

Objective 2 Find 10% and 1% of a number by moving the decimal point.

Video Examples

Review these examples for Objective 2:

2a. Find the exact answer to the question by moving the decimal point.

What is 10% of $674?

To find 10% of $674, divide $674 by 10. Do the division by moving the decimal point one place to the left.
10% of $674 = $67.40.

3. Find the exact answer to each question by moving the decimal point.

a. What is 1% of $947?

To find 1% of $947, divide $947 by 100. Do the division by moving the decimal point two places to the left.
1% of $947 = $9.47

b. Find 1% of 68.9 miles.

To find 1% of 68.9 miles, divide 68.9 by 100. Move the decimal point two places to the left.
1% of 68.9 = 0.689 mile

Now Try:

2a. Find the exact answer to the question by moving the decimal point.
What is 10% of $932?

_____

3. Find the exact answer to each question by moving the decimal point.
a. What is 1% of $549?

_____

b. Find 1% of 30.7 miles.

_____

Objective 2 Practice Exercises

For extra help, see Examples 2–3 on pages 532–533 of your text.

Find the exact answer to each question by moving the decimal point.

4. 1% of 400 homes is _____ . 4. _____

Name: Date:
Instructor: Section:

5. 10% of 4920 televisions is _____ .   5. _____

6. 1% of $98 is _____ .   6. _____

Objective 3   Solve basic percent problems using the percent equation.

Video Examples

Review these examples for Objective 3:

6a. Write and solve a percent equation to answer the question.

   126 credits is 70% of how many credits?

   Translate the sentence into an equation. Write the percent in decimal form.

   126 credits is 70% of how many credits?
   $$126 = 0.70 \cdot n$$

   Recall that 0.70 is equivalent to 0.7, so use 0.7 in the equation. To solve, divide both sides by 0.7.

   $$\frac{126}{0.7} = \frac{0.7 \cdot n}{0.7}$$

   $$180 = n$$

   So, 126 credits is 70% of 180 credits.

4a. Write and solve a percent equation to answer the question.

   35% of $245 is how much money?

   Translate the sentence into an equation, where *of* indicates multiplication and *is* translates to the equal sign. The percent must be written in decimal form.

   35% of $245 is how much money?
   $$0.35 \cdot 245 = n$$

   To solve, simplify the left side.
   $$(0.35)(245) = n$$
   $$85.75 = n$$

   So, 35% of $245 is $85.75.

Now Try:

6a. Write and solve a percent equation to answer the question.

   48 tons is 16% of how many tons?

   _____

4a. Write and solve a percent equation to answer the question.

   89% of $310 is how much money?

   _____

Name: Date:
Instructor: Section:

5a. Write and solve a percent equation to answer the question.

16 pounds is what percent of 200 pounds?

Translate the sentence into an equation. This time the percent is unknown.

16 pounds is what percent of 200 pounds?

$$16 = p \cdot 200$$

To solve the equation, divide both sides by 200.

$$\frac{16}{200} = \frac{p \cdot 200}{200}$$

$$0.08 = p$$

$$0.08 = 8\%$$

So 16 pounds is 8% of 200 pounds.

5a. Write and solve a percent equation to answer the question.

66 hours is what percent of 30 hours?

_____

Objective 3 Practice Exercises

For extra help, see Examples 4–6 on pages 534–537 of your text.

Write and solve an equation to answer each question.

7. 195 calls is what percent of 260 calls? 7. _____

8. 3% of $720 is how much? 8. _____

9. 24 magazines is 6% of what number of magazines? 9. _____

Name: Date:
Instructor: Section:

Chapter 7 PERCENT

7.4 Problem Solving with Percent

Learning Objectives
1 Solve percent application problems.
2 Solve problems involving percent of increase or decrease.

Key Terms

Use the vocabulary terms listed below to complete each statement in exercises 1–3.

percent of increase or decrease   **percent proportion**

percent equation

1. The statement part = percent · whole is called the _____.

2. In a _____ problem, the increase or decrease is expressed as a percent of the original amount.

3. The statement $\frac{\text{part}}{\text{whole}} = \frac{\text{percent}}{100}$ is called the _____.

Objective 1 Solve percent application problems.

Video Examples

Review these examples for Objective 1:

4. Janelle had budgeted $250 for new school clothes but ended up spending $390. The amount she spent was what percent of her budget?

   Step 1 Read the problem. It is about comparing her budget to the amount spent.
   Unknown: The percent of her budget
   Known: $250 budgeted, $390 spent.

   Step 2 Assign a variable. Let p be the unknown percent.

   Step 3 Write an equation.

   $\underbrace{\text{Amount spent}}_{390} \text{ is } \underbrace{\text{what percent}}_{p} \text{ of } \underbrace{\text{amount budgeted?}}_{250}$

   $390 = p \cdot 250$

Now Try:

4. Total Fitness Club predicted that 240 new members would join after Christmas. It actually had 396 new members join. The actual number joining is what percent of the predicted number?

   _____

Name: Date:
Instructor: Section:

Step 4 Solve the equation.
$$390 = p \cdot 250$$
$$\frac{390}{250} = \frac{p \cdot 250}{250}$$
$$1.56 = p$$
$$1.56 = 156\%$$

Step 5 State the answer. Janelle spent 156% of her budget.

Step 6 Check the solution. 100% is $250, and 50% is $125. So $250 + $125 = $350, or 150%, which is close to 156%.

5. David has 4.5% of his earnings deposited into a money market. If this amounts to $146.25 per month, find his monthly earnings.

   Step 1 Read the problem. It is about earnings.
   Unknown: monthly salary
   Known: $146.25 is 4.5% of monthly earnings.

   Step 2 Assign a variable. Let n = monthly salary.

   Step 3 Write an equation.

   4.5% of how much is $146.25

   $$0.045 \cdot n = 146.25$$

   Step 4 Solve the equation.
   $$0.045 \cdot n = 146.25$$
   $$\frac{0.045 \cdot n}{0.045} = \frac{146.25}{0.045}$$
   $$n = 3250$$

   Step 5 State the answer. David's monthly earnings is $3250.

   Step 6 Check the solution. Round 4.5% to 5%. 10% of $3250 is $325, then 5% is half of $325 or $162.50, which is close to the number given.

5. A multivitamin contains 10 micrograms of vitamin K. If this is 13% of the recommended daily dosage, what is the recommended daily dosage of vitamin K? Round the answer to the nearest whole number.

   _____

Name: Date:
Instructor: Section:

Objective 1 Practice Exercises

For extra help, see Examples 1–5 on pages 544–547 of your text.

Use the six problem-solving steps to answer each question. Round percent answers to the nearest tenth of a percent.

1. Members who are between 25 and 45 years of age make up 92% of the total membership of an organization. If there are 850 total members in the organization, find the number of members in the 25 to 45 age group.

    1. _____

2. Payroll deductions are 35% of Jason's gross pay. If his deductions total $350, what is his gross pay?

    2. _____

3. Vera's Antique Shoppe says that of its 5100 items in stock, 4233 are just plain junk, while the rest are antiques. What percent of the number of items in stock is antiques?

    3. _____

Name: Date:
Instructor: Section:

Objective 2   Solve problems involving percent of increase or decrease.

Video Examples

Review these examples for Objective 2:

6. Over the last three years, Calvin's salary has increased from $2700 per month to $3200. What is the percent increase?

    Step 1 Read the problem. It is about a salary increase.
        Unknown: percent of increase
        Known: original salary was $2700;
              new salary is $3200.

    Step 2 Assign a variable. Let p be the percent of increase.

    Step 3 Write an equation. First, subtract $3200 - $2700 to find the amount of increase.
        $3200 - $2700 = $500

    percent of original wage is amount of increase
    $$p \cdot 2700 = 500$$

    Step 4 Solve the equation.
    $$p \cdot 2700 = 500$$
    $$\frac{p \cdot 2700}{2700} = \frac{500}{2700}$$
    $$p \approx 0.185$$
    $$0.185 = 18.5\%$$

    Step 5 State the answer. Calvin's salary increased by 18.5%.

    Step 6 Check the solution. Round 18.5% to 20%. Then 10% of $2700 is $270, and 20% is two times 10%, or $540, which is close to the number given.

7. During the holiday season, average daily attendance at a health club fell from 495 members to 230 members. What was the percent decrease?

    Step 1 Read the problem. It is about decrease in attendance.
        Unknown: percent of decrease
        Known: original attendance: 495
              new attendance: 230.

Now Try:

6. A business increased the number of phone lines from 4 to 9. What is the percent increase?

    _____

7. The earnings per share of Amy's Cosmetic Company decreased from $1.20 to $0.86 in the last year. Find the percent of decrease.

    _____

Name: Date:
Instructor: Section:

Step 2 Assign a variable. Let p be the percent of decrease.

Step 3 Write an equation. First subtract 230 from 495 to find the amount of decrease.
$$495 - 230 = 265$$

$$\underbrace{\text{percent}}_{p} \text{ of } \underbrace{\text{original attendance}}_{495} \text{ is } \underbrace{\text{amount of decrease}}_{265}$$

$$p \cdot 495 = 265$$

Step 4 Solve the equation.
$$p \cdot 495 = 265$$
$$\frac{p \cdot 495}{495} = \frac{265}{495}$$
$$p \approx 0.535$$
$$0.535 = 53.5\%$$

Step 5 State the answer. The attendance decreased by 53.5%.

Step 6 Check the solution. 50% is half of 495, or 247.5, so 53.5% is a reasonable solution.

Objective 2 Practice Exercises

For extra help, see Examples 6–7 on pages 548–549 of your text.

Use the six problem-solving steps to find the percent increase or decrease. Round your answers to the nearest tenth of a percent.

4. Students at Withrow's College were charged $1560 for tuition this semester. If the tuition was $1480 last semester, find the percent of increase.

4. _____

Name: Date:
Instructor: Section:

5. Tomika's part-time work schedule has been reduced to 20 hours per week. She had been working 28 hours per week. What is the percent decrease?

5. _____

6. During a sale, the price of a futon was cut from $1250 to $999. Find the percent of decrease in price.

6. _____

Name: Date:
Instructor: Section:

Chapter 7 PERCENT

7.5 Consumer Applications: Sales Tax, Tips, Discounts, and Simple Interest

Learning Objectives
1 Find sales tax and total cost.
2 Estimate and calculate restaurant tips.
3 Find the discount and sale price.
4 Calculate simple interest and the total amount due on a loan.

Key Terms

Use the vocabulary terms listed below to complete each statement in exercises 1–8.

    sales tax   tax rate   discount

    interest   principal   interest rate

    simple interest   interest formula

1. The charge for money borrowed or loaned, expressed as a percent, is called _____.

2. A fee paid for borrowing or lending money is called _____.

3. The formula $I = p \cdot r \cdot t$ is the _____.

4. Use the formula $I = p \cdot r \cdot t$ to compute the amount of _____ due on a loan.

5. The amount of money borrowed or loaned is called the _____.

6. The percent of the total sales charged as tax is called the _____.

7. The percent of the original price that is deducted from the original price is called the _____.

8. The percent used when calculating the amount of tax is called the _____.

Objective 1 Find sales tax and total cost.

Video Examples

Review this example for Objective 1:

1. A television sells for $750 plus 8% sales tax. Find the price of the TV including sales tax.

   Step 1 Read the problem. It asks for the tax and the price of the TV.

Now Try:

1. A cell phone is $89. The sales tax is 7%. What is the tax and total?

   _____

Name: Date:
Instructor: Section:

Step 2 Assign a variable. Let *n* be the amount of tax.

Step 3 Write an equation. Use the sales tax equation.

$$\underbrace{\text{tax rate}}_{0.08} \cdot \underbrace{\text{cost of item}}_{750} = \underbrace{\text{sales tax}}_{n}$$

Step 4 Solve.
$$(0.08)(750) = n$$
$$60 = n$$

Add the sales tax to the cost of the TV to find the total cost.

$$\underbrace{\text{cost of item}}_{\$750} + \underbrace{\text{sales tax}}_{\$60} = \underbrace{\text{total cost}}_{\$810}$$

Step 5 State the answer. The tax is $60 and the total cost of the TV, including tax, is $810.

Step 6 Check. Use estimation to check that the amount of sales tax is reasonable. Round 8% to 10%. Then 10% of $750 is $75. It is reasonable.

Objective 1 Practice Exercises

For extra help, see Examples 1–2 on pages 554–555 of your text.

Find the amount of sales tax and the total cost. Round answers to the nearest cent, if necessary.

	Amount of sale	Tax Rate
1.	$350	6.5%

1. Tax _____

   Total _____

Find the sales tax rate. Round answers to the hundredth, if necessary.

	Amount of sale	Amount of Tax
2.	$450	$36

2. _____

| 3. | $215 | $10.75 |

3. _____

Name: Date:
Instructor: Section:

Objective 2 Estimate and calculate restaurant tips.

Video Examples

Review this example for Objective 2:

3a. First estimate the tip. Then calculate the exact answer.
Find a 15% tip for a restaurant bill of $87.22.

Estimate: Round $87.22 to $90. Then 10% of $90 is $9, and 5% is half of $9, or $4.50. So the estimate is $9 + $4.50 = $13.50.
Exact: Use the percent equation. Write 15% as a decimal. The bill for food and beverages is the whole and the tip is the part.

$$\text{percent} \cdot \text{whole} = \text{part}$$
$$15\% \cdot \$87.22 = n$$
$$(0.15)(87.22) = n$$
$$13.083 = n$$

Round $13.083 to $13.08 (nearest cent), which is close to the estimate of $13.50.

Now Try:

3a. First estimate the tip. Then calculate the exact answer.
Find a 15% tip for a restaurant bill of $23.75.

_____

Objective 2 Practice Exercises

For extra help, see Example 3 on pages 555–556 of your text.

For each restaurant bill, estimate a 15% tip and a 20% tip. Then find the exact amounts for a 15% tip and a 20% tip. Round exact amounts to the nearest cent if necessary.

4. $43.16

4. 15% estimate _____

15% exact _____

20% estimate _____

20% exact _____

5. $63.85

5. 15% estimate _____

15% exact _____

20% estimate _____

20% exact _____

Name: Date:
Instructor: Section:

6. $72.81

6. 15% estimate _____

15% exact _____

20% estimate _____

20% exact _____

Objective 3 Find the discount and sale price.

Video Examples

Review this example for Objective 3:
4. Mike Lee can purchase a new car at 8% below window sticker price. Find the sale price on a car with a window sticker price of $17,600.

Step 1 Read the problem. The problem asks for the price of the car after a discount of 8%.

Step 2 Work out a plan. The problem is solved in two steps. First, find the amount of the discount, that is, the amount that will be "taken off" (subtracted), by multiplying the original price ($17,600) by the rate of the discount (8%). The second step is to subtract the amount of discount from the original price. This gives the sale price, which is what Mike will actually pay for the car.

Step 3 Estimate a reasonable answer. Round the original price from $17,600 to $20,000, and the rate of discount from 8% to 10%. Since 10% is equivalent to $\frac{1}{10}$, the estimated discount is $20,000 ÷ 10 = 2000, so the estimated sale price is $20,000 − $2000 = $18,000.

Step 4 Solve the problem. First find the exact amount of the discount.
  amount of discount = rate of discount · original price
$$a = (8\%)(\$17,600)$$
$$a = (0.08)(\$17,600)$$
$$a = \$1408$$
Now, find the sale price of the car by subtracting the amount of the discount ($1408) from the original price.

Now Try:
4. A hard-cover book with an original price of $24.95 is on sale at 60% off. Find the sale price of the book.

_____

Name: Date:
Instructor: Section:

sale price = original price − amount of discount
= $17,600 − $1408
= $16,192

Step 5 State the answer. The sale price of the car is $16,192.

Step 6 Check. The exact answer, $16,192, is close to the estimate of $18,000.

Objective 3 Practice Exercises

For extra help, see Example 4 on page 557 of your text.

Find the amount of discount and the amount paid after the discount. Round money answers to the nearest cent, if necessary.

	Original price	**Rate of Discount**	
7.	$200	15%	7. Discount _____
			Amount paid _____

Solve each application problem. Round money answers to the nearest cent, if necessary.

8. A "Super 35% Off Sale" begins today. What is the price of a hair dryer normally priced at $15?

8. _____

9. Geishe's Shoes sells shoes at 33% off the regular price. Find the price of a pair of shoes normally priced at $54, after the discount is given.

9. _____

Name: Date:
Instructor: Section:

Objective 4 Calculate simple interest and the total amount due on a loan.

Video Examples

Review this example for Objective 4:
7. Find the simple interest and total amount due on $720 at $3\frac{1}{2}$% for 8 months.

   The principal is $720. The rate is $3\frac{1}{2}$% or 0.035 as a decimal, and the time is $\frac{8}{12}$ of a year. Use the formula $I = p \cdot r \cdot t$.

   $I = (720)(0.035)\left(\frac{8}{12}\right)$
   $= 25.2\left(\frac{2}{3}\right)$
   $= 16.80$

   The interest is $16.80.
   The total amount due is
   $720 + $16.80 = $736.80.

Now Try:
7. Find the simple interest and total amount due on $840 at $8\frac{1}{2}$% for 5 months.

   _____

Objective 4 Practice Exercises

For extra help, see Examples 5–7 on pages 558–559 of your text.

Find the simple interest and the total amount due. Round to the nearest cent, if necessary.

	Principal	Rate	Time in Years	
10.	$5280	9%	1	10. _____ _____
11.	$780	10%	$2\frac{1}{2}$	11. _____ _____
12.	$14,400	7%	7 months	12. _____ _____

Name: Date:
Instructor: Section:

Chapter 8 MEASUREMENT

8.1 Problem Solving with U.S. Measurement Units

Learning Objectives
1. Learn the basic U.S. measurement units.
2. Convert among U.S. measurement units using multiplication or division.
3. Convert among measurement units using unit fractions.
4. Solve application problems using U.S. measurement units.

Key Terms

Use the vocabulary terms listed below to complete each statement in exercises 1–3.

    U.S. measurement units   **unit fractions**   **metric system**

1. The _____ is based on units of ten.

2. _____ are used to convert among different measurements.

3. _____ include inches, feet, quarts, and pounds.

Objective 1   Learn the basic U.S. measurement units.

Video Examples

Review these examples for Objective 1:	Now Try:
1. Memorize the U.S. measurement conversions from the text. Then fill in the blanks.	1. Memorize the U.S. measurement conversions from the text. Then fill in the blanks.
a. 2 c = _____ pt	**a.** 16 oz = _____ lb
Answer: 1 pt	_____
b. 1 mi = _____ ft	**b.** 1 gal = _____ qt
Answer: 5280 ft	_____

Objective 1 Practice Exercises

For extra help, see Example 1 on page 476 of your text.

Fill in the blanks.

1. 1 T = _____ lb   1. _____

2. _____ pt = 1 qt   2. _____

3. 1 c = _____ fl oz   3. _____

Name: Date:
Instructor: Section:

Objective 2 Convert among U.S. measurement units using multiplication or division.

Video Examples

Review these examples for Objective 2:

2. Convert each measurement.

   a. 16 yd to feet

   You are converting from a larger unit to a smaller unit, so multiply.
   Because 1 yd = 3 ft, multiply by 3.
   $$16 \text{ yd} = 16 \cdot 3 = 48 \text{ ft}$$

   c. 8 pt to quarts

   You are converting from a smaller unit to a larger unit, so divide.
   Because 2 pt = 1 qt, divide by 2.
   $$8 \text{ pt} = \frac{8}{2} = 4 \text{ qt}$$

   d. 3960 ft to miles

   You are converting from a smaller unit to a larger unit, so divide.
   Because 5280 ft = 1 mi, divide by 5280.
   $$3960 \text{ ft} = \frac{3960}{5280} = \frac{3}{4} \text{ mi}$$

Now Try:

2. Convert each measurement.

   a. $17\frac{1}{2}$ ft to inches

   _____

   c. 75 sec to minutes

   _____

   d. 380 min to hours

   _____

Objective 2 Practice Exercises

For extra help, see Example 2 on page 477 of your text.

Convert each measurement using multiplication or division.

4. 12 ft to yards

4. _____

5. 40 pt to gallons

5. _____

6. 30 in to yards

6. _____

Name: Date:
Instructor: Section:

Objective 3 Convert among measurement units using unit fractions.

Video Examples

Review these examples for Objective 3:

3.
    a. Convert 64 oz to pounds.

Use a unit fraction with pounds (the unit for your answer) in the numerator, and ounces (the unit being changed) in the denominator. Because 1 lb = 16 oz, the necessary unit fraction is

$\dfrac{1 \text{ lb}}{16 \text{ oz}}$ ← Unit for your answer is pounds.
          ← Unit being changed is ounces.

Next, multiply 64 oz times this unit fraction. Write 64 oz as the fraction $\dfrac{64 \text{ oz}}{1}$ and divide out common units and factors wherever possible.

$$\dfrac{64 \text{ oz}}{1} \cdot \dfrac{1 \text{ lb}}{16 \text{ oz}} = \dfrac{\overset{4}{\cancel{64}} \ \cancel{\text{oz}}}{1} \cdot \dfrac{1 \text{ lb}}{\cancel{16} \ \cancel{\text{oz}}} = \dfrac{4 \cdot 1 \text{ lb}}{1} = 4 \text{ lb}$$

    b. Convert 8 lb to ounces.

Select the correct unit fraction to change 8 lb to ounces.

$\dfrac{16 \text{ oz}}{1 \text{ lb}}$ ← Unit for your answer is ounces.
        ← Unit being changed is pounds.

Multiply 8 lb times the unit fraction.

$$\dfrac{8 \text{ lb}}{1} \cdot \dfrac{16 \text{ oz}}{1 \text{ lb}} = \dfrac{8 \ \cancel{\text{lb}}}{1} \cdot \dfrac{16 \text{ oz}}{1 \ \cancel{\text{lb}}} = \dfrac{8 \cdot 16 \text{ oz}}{1} = 128 \text{ oz}$$

5b. Sometimes you may need to use two or three unit fractions to complete a conversion.

Convert 3 miles to inches.

Use three unit fractions. The first one changes from miles to yards, the next one changes yards to feet, and the last one changes feet to inches. All the units divide out except inches, which is the unit you want in your answer.

$$\dfrac{3 \ \cancel{\text{mi}}}{1} \cdot \dfrac{1760 \ \cancel{\text{yd}}}{1 \ \cancel{\text{mi}}} \cdot \dfrac{3 \ \cancel{\text{ft}}}{1 \ \cancel{\text{yd}}} \cdot \dfrac{12 \text{ in.}}{1 \ \cancel{\text{ft}}} = 190{,}080 \text{ inches}$$

Now Try:

3.
    a. Convert 12 yd to inches.

    _____

    b. Convert 4 in. to feet.

    _____

5b. Sometimes you may need to use two or three unit fractions to complete a conversion.
Convert 5 days to minutes.

    _____

Name: Date:
Instructor: Section:

Objective 3 Practice Exercises

For extra help, see Examples 3–5 on pages 478–480 of your text.

Convert each measurement using unit fractions.

7. 28 pt to gallons

7. _____

8. 38 c to pints

8. _____

9. 60 oz to pounds

9. _____

Objective 4 Solve application problems using U.S. measurement units.

Video Examples

Review this example for Objective 4:
6a. Answer the application problem.

Lee paid $4.65 for 14 oz of honey baked ham. What is the price per pound, to the nearest cent?

Step 1 Read the problem. The problem asks for the price per pound of the ham.

Step 2 Work out a plan. The weight of the ham is given in ounces, but the answer must be cost per pound. Convert ounces to pounds. The word per indicates division. You need to divide the cost by the number of pounds.

Step 3 Estimate a reasonable answer. To estimate, round $4.65 to $5, and round 14 ounces to 16 ounces. Then there are 16 oz in a pound, so there is about 1 pound. Finally, $5 ÷ 1 = $5 per pound is our estimate.

Step 4 Solve the problem. Use a unit fraction to convert 14 oz to pounds.

$$\frac{\cancel{14}^{7} \text{ oz}}{1} \cdot \frac{1 \text{ lb}}{\cancel{16}_{8} \text{ oz}} = \frac{7}{8} \text{ lb} = 0.875 \text{ lb}$$

Now Try:
6a. Answer the application problem.

Clarissa paid $1.79 for 4.5 oz of nuts. What is the cost per pound, to the nearest cent?

_____

Name: Date:
Instructor: Section:

Then divide to find the cost per pound.

$$\frac{\$4.65}{0.875} \approx \$5.31$$

Step 5 State the answer. The ham is $5.31 per pound (nearest cent).

Step 6 Check your work. The exact answer, $5.31 is close to the estimate of $5.

Objective 4 Practice Exercises

For extra help, see Example 6 on pages 480–481 of your text.

Solve each application problem using the six problem-solving steps.

10. Tony paid $2.84 for 6.5 oz of fudge. What is the cost per pound, to the nearest cent?

10. _____

11. At an office, each of 15 workers drinks about $1\frac{3}{4}$ c of coffee with each day. The office is open 4 days a week. How many quarts of coffee are needed for a week?

11. _____

Name: Date:
Instructor: Section:

Chapter 8 MEASUREMENT

8.2 The Metric System—Length

Learning Objectives
1. Learn the basic metric units of length.
2. Use unit fractions to convert among metric units.
3. Move the decimal point to convert among metric units.

Key Terms

Use the vocabulary terms listed below to complete each statement in exercises 1–3.

    meter   prefix   metric conversion line

1. Attaching a _____ such as "kilo-" or "milli" to the words "meter", "liter", or "gram" gives the names of larger or smaller units.

2. A line showing the various metric measurement prefixes and their size relationship to each other is called a _____.

3. The basic unit of length in the metric system is the _____.

Objective 1   Learn the basic metric units of length.

Video Examples

Review these examples for Objective 1:

1. Write the most reasonable metric unit in each blank. Choose from km, m, cm, and mm.

   a. The man's foot is 28 _____ .

   28 cm because cm are used instead of inches.
   28 cm is about 11 inches.

   b. The living room is 4 _____ .

   4 m because m are used instead of feet.
   4 m is about 13 feet.

   c. Sam rode his bike 8 _____ on the trail.

   8 km because km are used instead of miles.
   8 km is about 5 miles.

Now Try:

1. Write the most reasonable metric unit in each blank. Choose from km, m, cm, and mm.

   a. A strand of hair is 1 _____ wide.

   _____

   b. The truck sped down the highway at 120 _____ per hour.

   _____

   c. The width of a piece of paper is 21.6 _____ .

   _____

Name: Date:
Instructor: Section:

Objective 1 Practice Exercises

For extra help, see Example 1 on page 487 of your text.

Choose the most reasonable metric unit. Choose from **km**, **m**, **cm**, *or* **mm**.

1. the width of a twin bed   1. _____

2. the thickness of a dime   2. _____

3. the distance driven in 2 hours   3. _____

Objective 2 Use unit fractions to convert among metric units.

Video Examples

Review these examples for Objective 2:

2. Convert each measurement using unit fractions.

   a. 8 m to cm

   Put the unit for the answer (cm) in the numerator of the unit fraction; put the unit you want to change (m) in the denominator.

   $\frac{100 \text{ cm}}{1 \text{ m}}$ ← Unit for answer
   ← Unit being changed

   Multiply. Divide out common units where possible.

   $8m \cdot \frac{100cm}{1m} = \frac{8m}{1} \cdot \frac{100cm}{1m} = \frac{8 \cdot 100cm}{1} = 800cm$

   8 m = 800 cm

   b. 13.5 mm to cm

   Multiply by a unit fraction that allows you to divide out millimeters.

   $\frac{13.5 \text{ mm}}{1} \cdot \frac{1 \text{ cm}}{10 \text{ mm}} = \frac{13.5 \text{ cm}}{10} = 1.35 \text{ mm}$

   13.5 mm = 1.35 cm
   There are 10 mm in a cm, so 13.5 mm will be a smaller part of a cm.

Now Try:

2. Convert each measurement using unit fractions.

   a. 5400 m to km

   _____

   b. 7.6 cm to mm

   _____

Name: Date:
Instructor: Section:

Objective 2 Practice Exercises

For extra help, see Example 2 on page 488 of your text.

Convert each measurement using unit fractions.

  4. 25.87 m to centimeters   4. _____

  5. 450 m to kilometers   5. _____

  6. 140 millimeters to meters   6. _____

Objective 3 Move the decimal point to convert among metric units.

Video Examples

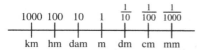

Review these examples for Objective 3:
3. Use the metric conversion line to make the following conversions.

   a. 62.892 km to m

   Find km on the metric conversion line. To get m, you move three places to the right. So move the decimal point in 62.892 three places to the right.
   62.892 km = 62,892 m

   b. 47.6 cm to m

   Find cm on the conversion line. To get m, move two places to the left. So move the decimal point two places to the left.
   47.6 cm = 0.476 m

   c. 52.8 mm to cm

   From mm to cm is one place to the left.
   52.8 mm = 5.28 cm

Now Try:
3. Use the metric conversion line to make the following conversions.
   a. 67.5 cm to mm

   _____

   b. 986.5 m to km

   _____

   c. 4.31 m to cm

   _____

254   Copyright © 2018 Pearson Education, Inc.

Name: Date:
Instructor: Section:

4a. Convert using the metric conversion line.

1.79 m to mm

Moving from m to mm is going three places to the right. In order to move the decimal point in 1.79 three places to the right, you must add a 0 as a place holder.
   1.790
1.79 m = 1790 mm

4a. Convert using the metric conversion line.
23 m to mm

_____

Objective 3 Practice Exercises

For extra help, see Examples 3–4 on pages 489–490 of your text.

Convert each measurement using the metric conversion line.

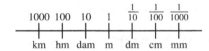

7.   1.94 cm to millimeters   7. _____

8.   10.35 km to meters   8. _____

9.   3.5 cm to kilometers   9. _____

Name: Date:
Instructor: Section:

Chapter 8 MEASUREMENT

8.3 The Metric System—Capacity and Weight (Mass)

Learning Objectives	
1	Learn the basic metric units of capacity.
2	Convert among metric capacity units.
3	Learn the basic metric units of weight (mass).
4	Convert among metric weight (mass) units.
5	Distinguish among basic metric units of length, capacity, and weight (mass).

Key Terms

Use the vocabulary terms listed below to complete each statement in exercises 1–2.

liter   **gram**

1. The basic unit of weight (mass) in the metric system is the _____.

2. The basic unit of capacity in the metric system is the _____.

Objective 1   Learn the basic metric units of capacity.

Objective 1 Practice Exercises

For extra help, see Example 1 on page 494 of your text.

Choose the most reasonable metric unit. Choose from **L,** *or* **ml**.

1. the amount of soda in a can   1. _____

2. the amount of orange juice in a large bottle   2. _____

3. the amount of cough syrup in one dose   3. _____

Objective 2   Convert among metric capacity units.

Video Examples

Review these examples for Objective 2:	Now Try:

2. Convert using the metric conversion line or unit fractions.
   a. 5.4 L to mL
 	
Using the metric conversion line:
From L to mL is three places to the right.
   5.4 L = 5400 mL
 	
Using unit fractions:
Multiply by a unit fraction that allows you to
	2. Convert using the metric conversion line or unit fractions.
	   a. 973 mL to L
	
	_____

Name: Date:
Instructor: Section:

divide out liters.

$$\frac{5.4 \cancel{L}}{1} \cdot \frac{1000 \text{ mL}}{1 \cancel{L}} = 5400 \text{ mL}$$

b. 92 mL to L

Using the metric conversion line:
From mL to L is three places to the left.
92 mL = 0.092 L

Using unit fractions:
Multiply by a unit fraction that allows you to divide out mL.

$$\frac{9.2 \cancel{\text{mL}}}{1} \cdot \frac{1 \text{ L}}{1000 \cancel{\text{mL}}} = 0.092 \text{ L}$$

b. 3.85 L to mL

_____

Objective 2 Practice Exercises

For extra help, see Example 2 on pages 494–495 of your text.

Convert each measurement. Use unit fractions or the metric conversion line.

4. 2.5 L to milliliters   4. _____

5. 836 kL to liters   5. _____

6. 7863 mL to liters   6. _____

Objective 3   Learn the basic metric units of weight (mass).

Video Examples

Review these examples for Objective 3:
3. Write the most reasonable metric unit in each blank. Choose from kg, g, and mg.

   a. The watermelon weighed 18 _____ .

   18 kg because kilograms are used instead of pounds.
   18 kg is about 40 pounds.

Now Try:
3. Write the most reasonable metric unit in each blank. Choose from kg, g, and mg.
   a. A fashion model weighs 48 _____ .

   _____

Name: Date:
Instructor: Section:

b. The vitamin tablet was 400 _____ .

400 mg because 400 g would be more than two hamburgers, which is too much.

c. A birthday card weighs 8 _____ .

8 grams because 8 kg would be too heavy.

b. Jacob's football weighs 590 _____ .

_____

c. The pencil lead weighs 3 _____ .

_____

Objective 3 Practice Exercises

For extra help, see Example 3 on page 496 of your text.

Choose the most reasonable metric unit. Choose from **kg**, **g**, *or* **mg**.

7. the weight (mass) of a grain of rice   7. _____

8. the weight (mass) of a car   8. _____

9. the weight (mass) of an egg   9. _____

Objective 4   Convert among metric weight (mass) units.

Video Examples

Review these examples for Objective 4:
4. Convert using the metric conversion line or unit fractions.

   a. 26 mg to g

   Using the metric conversion line:
   From mg to g is three places to the left.
   26 mg = 0.026 g

   Using unit fractions:
   Multiply by a unit fraction that allows you to divide out mg.

   $$\frac{26 \text{ mg}}{1} \cdot \frac{1 \text{ g}}{1000 \text{ mg}} = 0.026 \text{ g}$$

Now Try:
4. Convert using the metric conversion line or unit fractions.

   a. 3.72 g to mg

   _____

Name: Date:
Instructor: Section:

b. 97.34 kg to g

Using the metric conversion line:
From kg to g is three places to the right.
97.34 kg = 97,340 g

Using unit fractions:
Multiply by a unit fraction that allows you to divide out kg.

$$\frac{97.34 \cancel{\text{kg}}}{1} \cdot \frac{1000 \text{ g}}{1 \cancel{\text{kg}}} = 97,340 \text{ g}$$

b. 84 g to kg

_____

Objective 4 Practice Exercises

For extra help, see Example 4 on page 497 of your text.

Convert each measurement. Use unit fractions or the metric conversion line.

10. 27,000 g to kilograms   10. _____

11. 0.76 kg to grams   11. _____

12. 4.7 g to milligrams   12. _____

Objective 5   Distinguish among basic metric units of length, capacity, and weight (mass).

Video Examples

Review these examples for Objective 5:

5. First decide which type of unit is needed: length, capacity, or weight. Then write the most appropriate metric unit in the blank. Choose from km, m, cm, mm, L, mL, kg, g, and mg.

a. The tree is 9 _____ high.

Use length units because of the word high.
The tree is 9 m high.

Now Try:

5. First decide which type of unit is needed: length, capacity, or weight. Then write the most appropriate metric unit in the blank. Choose from km, m, cm, mm, L, mL, kg, g, and mg.

a. The moisturizer jar has 125 _____ .

_____

Name: Date:
Instructor: Section:

b. A ten-carat diamond weighs 2 _____ .

Use weight units because of the word weigh.
The ten-carat diamond weighs 2 g.

b. This is a 1.19 _____ box of breakfast cereal.

_____

Objective 5 Practice Exercises

For extra help, see Example 5 on page 498 of your text.

*Choose the most reasonable metric unit. Choose from **km**, **m**, **cm**, **mm, mL**, **L**, **kg**, **g**, or **mg**.*

13. Buy a 5 _____ bottle of water. 13. _____

14. The piece of wood weighs 5 _____ . 14. _____

15. A paperclip is 3 _____ long. 15. _____

Name: Date:
Instructor: Section:

Chapter 8 MEASUREMENT

8.4   Problem Solving with Metric Measurement

Learning Objectives
1   Solve application problems involving metric measurements.

Key Terms

Use the vocabulary terms listed below to complete each statement in exercises 1–3.

   meter   liter   gram

1. A _____ is the weight of 1 mL of water.

2. A _____ is a little longer than a yard.

3. A _____ is a little more than one quart.

Objective 1   Solve application problems involving metric measurements.

Video Examples

Review these examples for Objective 1:

2. A 70-L drum is filled with oil which is to be packaged into 140-mL bottles. How many bottles can be filled.?

   Step 1 Read the problem. The problem asks for the number of filled bottles.

   Step 2 Work out a plan. The given amount is in liters, but the capacity of the bottles is in mL. Convert liters to milliliters, then divide by 140 mL (the capacity of the bottles).

   Step 3 Estimate a reasonable answer. To estimate round 140 mL to 100 mL. Then 70 L = 70,000 mL, and $70,000 \div 100 = 700$ bottles.

   Step 4 Solve the problem. On the metric conversion line, moving from L to mL is three places to the right, so move the decimal point in 70 L three places to the right. Then divide by 140.

   70 L = 70,000 mL

   $$\frac{70,000 \text{ mL}}{140 \text{ mL}} = 500 \text{ bottles}$$

   Step 5 State the answer. 500 bottles will be filled.

Now Try:

2. If 1.8 kg of candy is to be divided equally among 9 children, how many grams will each child receive?

   _____

Name: Date:
Instructor: Section:

Step 6 Check your work. The exact answer of 500 bottles is close to our estimate of 700 bottles.

3. Lucy purchased 1 m 60 cm of fabric at $6.25 per meter for a jacket and 1m 80 cm of fabric at $4.75 per m for a dress. How much did she spend in total?

   Step 1 Read the problem. Lucy purchased two quantities of two different kinds of fabric. Find the total she spent on fabric.

   Step 2 Work out a plan. The lengths involve two units, m and cm. Rewrite both lengths in meters, determine the price, and find the total.

   Step 3 Estimate a reasonable answer. To estimate, 1 m 60 cm can be rounded to 2 m. Round 1 m 80 cm to 2 m. Also, $6.25 can be rounded to $6 and $4.75 can be rounded to $5. Then, $2 \times \$6 + 2 \times \$5 = \$12 + \$10 = \$22$, our estimate.

   Step 4 Solve the problem. Rewrite the lengths in meters.

   $$\begin{array}{ll} 1\text{ m} \rightarrow 1.0\text{ m} & 1\text{ m} \rightarrow 1.0\text{ m} \\ \text{plus }60\text{ cm} \rightarrow \underline{+\,0.6\text{ m}} & \text{plus }80\text{ cm} \rightarrow \underline{+\,0.8\text{ m}} \\ \phantom{\text{plus }60\text{ cm} \rightarrow }1.6\text{ m} & \phantom{\text{plus }80\text{ cm} \rightarrow }1.8\text{ m} \end{array}$$

   Jacket: $1.6 \times \$6.25 = \$10$
   Dress: $1.8 \times \$4.75 = \underline{\$\ 8.55}$
                             $18.55 total

   Step 5 State the answer. The total spent is $18.55.

   Step 6 Check your work. The exact answer of $18.55 is close to our estimate of $22.

3. A fish tank can hold up to 75.6 L of water. If there are 70,000 mL of water in the tank, how many more milliliters of water can the tank hold?

   _____

Name: Date:
Instructor: Section:

Objective 1 Practice Exercises

For extra help, see Examples 1–3 on pages 505–506 of your text.

Solve each application problem. Round money answers to the nearest cent.

1. Metal chain costs $5.26 per meter. Find the cost of 2 m 47 cm of the chain. Round your answer to the nearest cent.

   1. _____

2. Henry bowls with a 7-kg bowling ball, while Denise uses a ball that weighs 5 kg 750g. How much heavier is Henry's bowling ball than Denise's?

   2. _____

3. The label on a bottle of pills says that there are 3.5 mg of the medication in 5 pills. If a patient needs to take 8.4 mg of the medication, how many pills does he need to take?

   3. _____

Name: Date:
Instructor: Section:

Chapter 8 MEASUREMENT

8.5 Metric–U.S. Measurement Conversions and Temperature

Learning Objectives	
1	Use unit fractions to convert between metric and U.S. measurement units.
2	Learn common temperatures on the Celsius scale.
3	Use formulas to convert between Celsius and Fahrenheit temperatures.

Key Terms

Use the vocabulary terms listed below to complete each statement in exercises 1–2.

Celsius   **Fahrenheit**

1. The _____ scale is used to measure temperature in the metric system.

2. The _____ scale is used to measure temperature in the U.S. customary system.

Objective 1 Use unit fractions to convert between metric and U.S. measurement units.

Video Examples

Review these example for Objective 1:	Now Try:
1. Convert from 16 in. to centimeters.	1. Convert from 12.9 m to feet.
We're changing from a U.S length unit to a metric length unit. In the "U.S. to Metric Units" side of the table, you see that 1 inch $\approx$ 2.54 centimeters. Two unit fractions can be written using that information. $\dfrac{1 \text{ in.}}{2.54 \text{ cm}}$ or $\dfrac{2.54 \text{ cm}}{1 \text{ in.}}$ Multiply by the unit fraction that allows you to divide out inches. $16 \text{ in.} \cdot \dfrac{2.54 \text{ cm}}{1 \text{ in.}} = \dfrac{16 \cancel{\text{ in.}}}{1} \cdot \dfrac{2.54 \text{ cm}}{1 \cancel{\text{ in.}}} = 40.64 \text{ cm}$ $16 \text{ in.} \approx 40.64 \text{ cm}$	_____

Name: _____ Date: _____
Instructor: _____ Section: _____

2. Convert using unit fractions. Round your answers to the nearest tenth.

   a. 22.5 pounds to kilograms

   Look in the "U.S. to Metric Units" side of the table to see that 1 pound ≈ 0.45 kilograms. Use this information to write a unit fraction that allows you to divide out pounds.

   $$\frac{22.5 \cancel{\text{lb}}}{1} \cdot \frac{0.45 \text{ kg}}{1 \cancel{\text{lb}}} = 10.125 \text{ kg}$$

   22.5 lb ≈ 10.1 kg

   b. 46.8 L to quarts

   Look in the "Metric to U.S. Units" side of the table to see that 1 L ≈ 1.06 quarts. Write a unit fraction that allows you to divide out liters.

   $$\frac{46.8 \cancel{L}}{1} \cdot \frac{1.06 \text{ qt}}{1 \cancel{L}} = 49.608 \text{ qt}$$

   46.8 L ≈ 49.6 qt

2. Convert using unit fractions. Round your answers to the nearest tenth.

   a. 9.68 kg to pounds

   _____

   b. 20 gallons to liters

   _____

Objective 1 Practice Exercises

For extra help, see Examples 1–2 on pages 509–510 of your text.

Use the table in your textbook and unit fractions to make the following conversions. Round answers to the nearest tenth.

1. 291 mi to kilometers

   1. _____

2. 7 L to gallons

   2. _____

3. 26 oz to grams

   3. _____

Name: Date:
Instructor: Section:

Objective 2   Learn common temperatures on the Celsius scale.
Objective 2 Practice Exercises

For extra help, see Example 3 on page 511 of your text.

Choose the most reasonable temperature for each situation.

4. hot coffee
    35°C   60°C   100°C

    4. _____

5. Normal body temperature
    37°C   37°F

    5. _____

6. Oven temperature
    300°C   300°F

    6. _____

Objective 3   Use formulas to convert between Celsius and Fahrenheit temperatures.

Video Examples

Review these examples for Objective 3:

4. Convert 77°F to Celsius.

    Use the formula and follow the order of operations.

    $$C = \frac{5(F-32)}{9}$$
    $$= \frac{5(77-32)}{9}$$
    $$= \frac{5(45)}{9}$$
    $$= \frac{5(\cancel{45}^{5})}{\cancel{9}_{1}}$$
    $$= 25$$

    Thus, 77°F = 25°C.

Now Try:

4. Convert 122°F to Celsius.

    _____

Name: Date:
Instructor: Section:

5. Convert 30°C to Fahrenheit.

Use the formula and follow the order of operations.

$$F = \frac{9 \cdot C}{5} + 32$$
$$= \frac{9 \cdot 30}{5} + 32$$
$$= \frac{9 \cdot \overset{6}{\cancel{30}}}{\underset{1}{\cancel{5}}} + 32$$
$$= 54 + 32$$
$$= 86$$

Thus, 30°C = 86°F.

5. Convert 150°C to Fahrenheit.

_____

Objective 3 Practice Exercises

For extra help, see Examples 4–5 on page 512 of your text.

Use the conversions formulas and the order of operations to convert Fahrenheit temperatures to Celsius and Celsius temperatures to Fahrenheit. Round your answers to the nearest degree, if necessary.

7. 62°F

7. _____

8. 10°C

8. _____

Solve the application problem. Round to the nearest degree, if necessary.

9. A recipe for roast beef calls for an oven temperature of 400°F. What is the temperature in degrees Celsius?

9. _____

Name: Date:
Instructor: Section:

Chapter 9 GRAPHS AND GRAPHING

9.1 Problem Solving with Tables and Pictographs

Learning Objectives
1. Read and interpret data presented in a table.
2. Read and interpret data from a pictograph.

Key Terms

Use the vocabulary terms listed below to complete each statement in exercises 1–2.

table   pictograph

1. A graph that uses pictures or symbols to display information is called a
   _____.

2. A display of facts in rows and columns is called a _____.

Objective 1   Read and interpret data presented in a table.

Video Examples

The table below shows information about the performance of the eight U.S. airlines during the year 2014 (January–December).

PERFORMANCE DATA FOR SELECTED U.S. AIRLINES
January–December 2014   *Luggage problems per 1000 passengers

Airline	On-Time Performance	Luggage Handling*
Alaska	80%	2.7
Delta	89%	2.3
Frontier	68%	1.8
Hawaiian	88%	2.2
Jetblue	80%	2.0
Skywest	67%	4.7
United	72%	3.7
Virgin America	68%	1.0

Review these examples for Objective 1:
1. Use the table above to answer these questions.

   a. What percent of Hawaiian's flights were on time?

   Look across the row labeled Hawaiian. 88% of its flights were on time.

Now Try:
1. Use the table above to answer these questions.

   a. What percent of Frontier's flights were on time?

   _____

Name: Date:
Instructor: Section:

b. Which airline had the best luggage handling record?

Look down the column headed Luggage Handling. To find the best record, look for the lowest number, which is 1.0. Then look to the left to find the airline, which is Virgin America.

b. What airline(s) had the second worst luggage handling record?

_____

The table below shows the maximum cab fares in five different cities in 2015. The "flag drop" change is made when the driver starts the meter. "Wait time" is the charge for having to wait in the middle of a ride.

MAXIMUM TAXICAB FARES ALLOWED IN SELECTED CITIES IN 2015

City	Flag Drop	Price per Mile	Wait Time (per Hour)
Chicago	$4.00	$2.00	$20
Houston	$2.75	$2.20	$24
Miami	$2.95	$5.10	$24
New York	$2.50	$2.50	$24
San Francisco	$3.50	$2.75	$32

Source: taxifarefinder.com

2. Use the table above to answer these questions.

a. What is the maximum fare for a 15-mile ride in Chicago that includes having the cab wait 12 minutes?

The price per mile in Chicago is $2, so the cost for 15 miles is 15($2) = $30.00. Then add the flag drop charge of $4. Finally, figure out the cost of the wait time. One way is to set up a proportion.

$$\text{Cost} \rightarrow \frac{\$20}{60 \text{ min}} = \frac{\$x}{12 \text{ min}} \leftarrow \text{Cost}$$
$$\text{Wait time} \rightarrow \qquad\qquad\qquad\qquad \leftarrow \text{Wait time}$$

$$60 \cdot x = 20 \cdot 12$$
$$\frac{60x}{60} = \frac{240}{60}$$
$$x = \$4$$

Total fare = $30 + $4 + $4 = $38.

2. Use the table above to answer these questions.

a. What is the maximum fare for a 8-mile ride in New York that includes having the cab wait 10 minutes?

_____

270 Copyright © 2018 Pearson Education, Inc.

Name: _____ Date: _____
Instructor: _____ Section: _____

b. It is customary to give the cab driver a tip. Find the total cost of the cab ride in part (a) if the passenger added at 10% tip, rounded to the nearest quarter (nearest $0.25).

Use the percent equation to find the exact tip.
percent · whole = part
$$10\%(\$38) = n$$
$$0.10(\$38) = n$$
$$\$3.80 = n \quad \text{Round to } \$3.75$$
$$\text{(nearest } \$0.25)$$

The total cost of the cab ride is $38 + $3.75 tip = $41.75.

b. It is customary to give the cab driver a tip. Find the total cost of the cab ride in part (a) if the passenger added at 20% tip, rounded to the nearest quarter (nearest $0.25).

_____

Objective 1 Practice Exercises

For extra help, see Examples 1–2 on pages 638–639 of your text.

Use the table below for exercises 1–3. Use proportions to help solve exercise 3.

Weight of Exerciser	123 lbs	130 lbs	143 lbs
	Calories burned in 30 minutes		
Cycling	168	177	195
Running	324	342	375
Jumping Rope	273	288	315
Walking	162	171	189

Source: Fitness magazine

1. How many calories are burned by a 130-pound adult in 30 minutes of cycling?

   1. _____

2. Which activity will burn at least 300 calories when performed by a 123-pound adult?

   2. _____

3. How many calories will a 123-pound adult burn in 60 minutes of walking and 15 minutes of running?

   3. _____

Copyright © 2018 Pearson Education, Inc.

Name: Date:
Instructor: Section:

Objective 2 Read and interpret data from a pictograph.

Video Examples

This pictograph shows the approximate number of passenger arrivals and departures at selected U.S. airports in 2006.

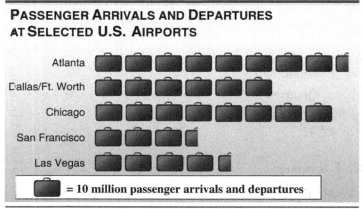

Review these examples for Objective 2:

3. Use the pictograph to answer these questions.

   a. Approximately how many passenger arrivals and departures took place at the San Francisco airport?

   The passenger arrivals and departures for San Francisco shows 3 whole symbols ($3 \cdot 10$ million = 30 million) plus half of a symbol ($\frac{1}{2}$ of 10 million is 5 million) for a total of 35 million.

   b. What is the difference in the number of arrivals and departures at Dallas/Ft. Worth airport and Atlanta airport?

   Dallas/Fort Worth shows 6 symbols and Atlanta shows $8\frac{1}{2}$ symbols. So Atlanta has $2\frac{1}{2}$ more symbols than Dallas/Fort Worth, and $2\frac{1}{2} \cdot 10$ million $= 25$ million. Thus, Atlanta has 25 million more passenger arrivals and departures than Dallas/Fort Worth.

Now Try:

3. Use the pictograph to answer these questions.

   a. Approximately how many passenger arrivals and departures took place at the Las Vegas airport?

   _____

   b. What is the difference in the number of arrivals and departures at Chicago airport and San Francisco airport?

   _____

Name: Date:
Instructor: Section:

Objective 2 Practice Exercises

For extra help, see Example 3 on page 640 of your text.

The pictograph below shows the population of five cities in 2010. Use the pictograph to answer exercises 4−6.

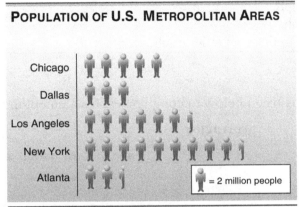

Source: U.S. Bureau of the Census, U.S. Department of Commerce.

4. What is the approximate population of Dallas? 4. _____

5. Approximately how much greater is the population of Los Angeles than the population of Chicago? 5. _____

6. What is the approximate total population of all five cities? 6. _____

Name: Date:
Instructor: Section:

Chapter 9 GRAPHS AND GRAPHING

9.2 Reading and Constructing Circle Graphs

Learning Objectives
1. Read a circle graph.
2. Use a circle graph.
3. Use a protractor to draw a circle graph.

Key Terms

Use the vocabulary terms listed below to complete each statement in exercises 1–2.

    circle graph   protractor

1. A _____ shows how a total amount is divided into parts or sectors.

2. A _____ is a device used to measure the number of degrees in angles or parts of a circle.

Objective 1   Read a circle graph.

Objective 1 Practice Exercises

For extra help, see page 645 of your text.

The circle graph shows the cost of remodeling a kitchen. Use the graph to answer exercises 1–3.

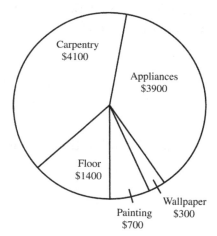

1. Find the total cost of remodeling the kitchen.   1. _____

2. What is the largest single expense in remodeling the kitchen?   2. _____

3. How much less does the wallpaper cost than painting?   3. _____

Name: Date:
Instructor: Section:

Objective 2 Use a circle graph.

Video Examples

The circle graph shows the expenses involved in keeping a sales force on the road. Each expense item is expressed as a percent of the total sales force cost of $950,000.

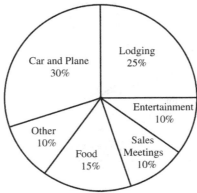

Review this example for Objective 2:

3. Use the circle graph above on expenses to find the amount spent on lodging.

   Recall the percent equation.
   $$\text{percent} \cdot \text{whole} = \text{part}$$
   The total expenses is $950,000, so the whole is $950,000. The percent is 25%, or as a decimal, 0.25. Find the part.
   $$\text{percent} \cdot \text{whole} = \text{part}$$
   $$(0.25)(950,000) = n$$
   $$237,500 = n$$
   The amount spent on lodging was $237,500.

Now Try:

3. Use the circle graph above on expenses to find the amount spent on food.

   _____

Name: Date:
Instructor: Section:

Objective 2 Practice Exercises

For extra help, see Examples 1–3 on pages 645–646 of your text.

The circle graph shows the number of students enrolled in certain majors at a college. Use the graph to answer exercise 4. The entire circle represents 11,600 students.

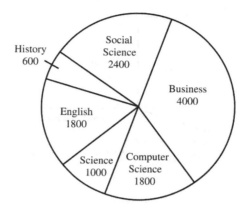

4. Find the ratio of the number of business majors to the total number of students.

4. _____

The circle graph shows the expenses involved in keeping a sales force on the road. Each expense item is expressed as a percent of the total sales force cost of $950,000. Find the number of dollars of expense for each category.

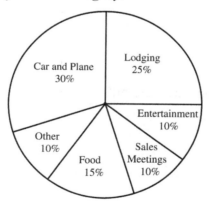

5. Entertainment

5. _____

6. Car and plane

6. _____

276 Copyright © 2018 Pearson Education, Inc.

Name: Date:
Instructor: Section:

Objective 3 Use a protractor to draw a circle graph.

Video Examples

Review these examples for Objective 3:

4. A family recorded its expenses for a year, with the following results: 40% for Housing, 20% for Food, 14% for Automobile, 8% for Clothing, 6% for Medical, 8% for Savings, and 4% for Other.

   a. Find the number of degrees in a circle graph for each type of expenses.

   Recall that a complete circle has 360°. Because "Housing" makes up 40% of the expenses, the number of degrees needed for the "Housing" sector of the circle graph is 40% of 360°.

   Housing
   $(360°)(40\%) = (360°)(0.40) = 144°$
   Food
   $(360°)(20\%) = (360°)(0.20) = 72°$
   Automobile
   $(360°)(14\%) = (360°)(0.14) = 50.4°$
   Clothing
   $(360°)(8\%) = (360°)(0.08) = 28.8$
   Medical
   $(360°)(6\%) = (360°)(0.06) = 21.6°$
   Savings
   $(360°)(8\%) = (360°)(0.08) = 28.8°$
   Other
   $(360°)(4\%) = (360°)(0.04) = 14.4°$

   b. Draw a circle graph showing this information.

   Use a protractor to make the circle graph.

   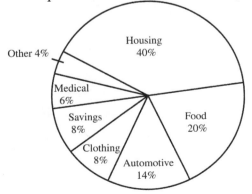

Now Try:

4. A book publisher had 30% of its sales in mysteries, 15% in biographies, 10% in cookbooks, 25% in romance novels, 15% in science, and the rest in business books.

   a. Find the number of degrees in a circle graph for each type of book.

   mysteries _____

   biographies _____

   cookbooks _____

   romance _____

   science _____

   business _____

   b. Draw a circle graph showing this information.

   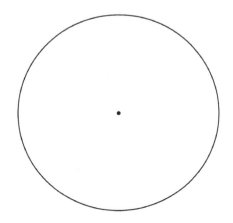

Name: Date:
Instructor: Section:

Objective 3 Practice Exercises

For extra help, see Example 4 on pages 647–648 of your text.

Use the given information to draw a circle graph.

Jensen Manufacturing Company has its annual sales divided into five categories as follows. The total sales for a year is $400,000.

Item	Annual Sales
Parts	$20,000
Hand tools	80,000
Bench tools	100,000
Brass fittings	140,000
Cabinet hardware	60,000

7. Find the percent of the total sales for each item.

7. parts _____

hand tools _____

bench tools _____

brass fittings _____

hardware _____

8. Find the number of degrees in a circle graph for each item.

8. parts _____

hand tools _____

bench tools _____

brass fittings _____

hardware _____

Name: Date:
Instructor: Section:

9. Make a circle graph showing this information.

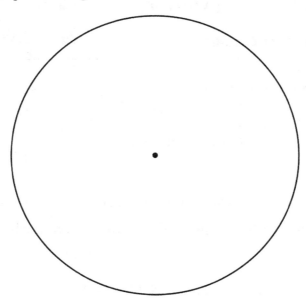

Name: Date:
Instructor: Section:

Chapter 9 GRAPHS AND GRAPHING

9.3 Bar Graphs and Line Graphs

Learning Objectives
1. Read and understand a bar graph.
2. Read and understand a double-bar graph.
3. Read and understand a line graph.
4. Read and understand a comparison line graph.

Key Terms

Use the vocabulary terms listed below to complete each statement in exercises 1–4.

bar graph   double-bar graph   line graph   comparison line graph

1. A _____ uses dots connected by a line to show trends.

2. A _____ compares two sets of data by showing two sets of bars.

3. A _____ uses bars of various heights or lengths to show quantity or frequency.

4. A _____ shows how two sets of data relate to each other by showing a line graph for each item.

Objective 1 Read and understand a bar graph.

Video Examples

The bar graph shows the enrollment for the fall semester at a small college from 2012 to 2016.

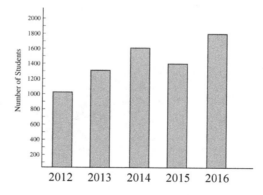

Review this example for Objective 1:
1. What was the fall enrollment for 2013?

   The bar for 2013 rises to 1300. So the enrollment in 2013 was 1300 students.

Now Try:
1. What was the fall enrollment for 2016?

   _____

280 Copyright © 2018 Pearson Education, Inc.

Name: Date:
Instructor: Section:

Objective 1 Practice Exercises

For extra help, see Example 1 on page 654 of your text.

The bar graph shows the enrollment for the fall semester at a small college from 2012 to 2016. Use this graph for exercises 1–3.

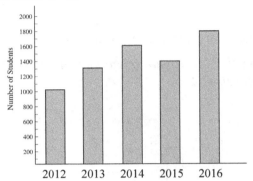

1. What was the fall enrollment for 2014?   1. _____

2. What year had the greatest enrollment?   2. _____

3. By how many students did the enrollment increase from 2015 to 2016?   3. _____

Objective 2   Read and understand a double-bar graph.

Video Examples

The double-bar graph shows the enrollment by gender in each class at a small college.

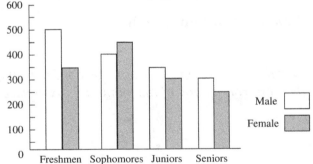

Review these examples for Objective 2:
2. Use the double-bar graph to find the following.

   a. The number of male sophomores enrolled

   There are two bars for the sophomore class. The color code to the right tells you that white bars represent male enrollments. So the white bar on the left for Sophomores represents the number of male sophomores. It rises to 400.
   So the sophomore enrollment of males is 400.

Now Try:
2. Use the double-bar above to find the following.
   a. The number of female seniors enrolled

   _____

b. The number of female juniors enrolled

The gray column for juniors rises to 300.
So the junior enrollment of females is 300.

b. The number of male freshmen enrolled

_____

Objective 2 Practice Exercises

For extra help, see Example 2 on page 655 of your text.

The double-bar graph shows the enrollment by gender in each class at a small college. Use the double-bar graph for exercises 4–6.

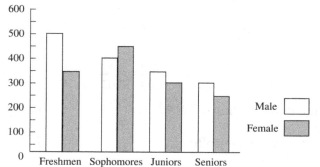

4. How many female freshmen are enrolled?   4. _____

5. Which class has a greater female enrollment than male enrollment?   5. _____

6. Find the total number of juniors enrolled.   6. _____

Objective 3   Read and understand a line graph.

Video Examples

The line graph gives the value of one share of stock of Microchip Computer Corporation on the first trading day of the month for six consecutive months.

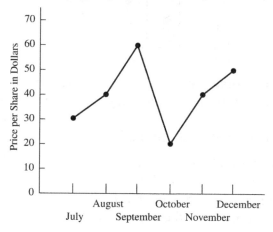

Name: Date:
Instructor: Section:

Review these examples for Objective 3:
3. Use the line graph on the previous page to find the following.

   a. In which month was the value of the stock the lowest?

   The lowest point on the graph is the dot directly over October, so the lowest value of the stock occurred in October.

   b. Find the value of one share of stock on the first trading day in August.

   Use a ruler or straightedge to line up the August dot with the numbers along the left edge of the graph. The August dot is directly across from 40. So in August, the price per share was $40.

Now Try:
3. Use the line graph on the previous page to find the following.

   a. Find the value of the stock in September.

   _____

   b. In which month was the value of the stock $50?

   _____

Objective 3 Practice Exercises

For extra help, see Example 3 on page 655 of your text.

The line graph gives the value of one share of stock of Microchip Computer Corporation on the first trading day of the month for six consecutive months. Use the line graph for exercises 7–9.

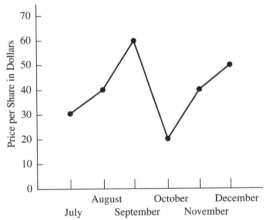

7. In which month was the value of the stock highest? 7. _____

8. Find the value of one share on the first trading day October. 8. _____

9. By how much did the value of one share increase from July to September? 9. _____

Name: Date:
Instructor: Section:

Objective 4 Read and understand a comparison line graph.

Video Examples

The comparison line graph shows annual sales for two different stores from 2012 to 2016.

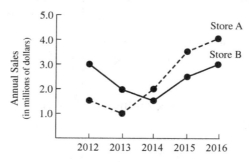

Review these examples for Objective 4:

4. Use the comparison line graph to find the following.

   a. The number of annual sales for store A in 2014

   Find the dot on the dotted line above 2004. It rises to $2.0. Multiply 2.0 by $1,000,000 because the label on the left side of the graph says in millions of dollars.
   So, in 2014 the annual sales for store A was $2,000,000.

   b. The number of annual sales for store B in 2016

   The solid line on the graph shows 3.0 times $1,000,000 or $3,000,000 was the annual sales for store B in 2016.

Now Try:

4. Use the comparison line graph to find the following.

   a. The number of annual sales for store B in 2014

   _____

   b. The number of annual sales for store A in 2016

   _____

Name: Date:
Instructor: Section:

Objective 4 Practice Exercises

For extra help, see Example 4 on page 656 of your text.

The comparison line graph shows annual sales for two different stores from 2012 to 2016. Use the graph to solve exercises 10–12.

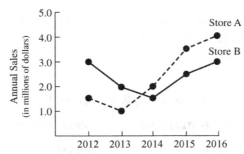

10. Find the annual sales for store A in 2015. 10. _____

11. Find the annual sales for store B in 2013. 11. _____

12. Find the amount of decrease and the percent of decrease in Store A's annual sales from 2012 to 2013. Round to the nearest whole percent, if necessary. 12. _____

Name: Date:
Instructor: Section:

Chapter 9 GRAPHS AND GRAPHING

9.4 The Rectangular Coordinate System

Learning Objectives
1. Plot a point, given the coordinates, and find the coordinates, given a point.
2. Identify the four quadrants and determine which points lie within each one.

Key Terms

Use the vocabulary terms listed below to complete each statement in exercises 1−8.

paired data	horizontal axis	vertical axis	
ordered pair	x-axis	y-axis	
coordinate system	origin	coordinates	quadrants

1. An _____ is the "address" of a point in a coordinate system.

2. In a rectangular coordinate system, the axis that goes "left and right" is called the _____ or the _____.

3. In a rectangular coordinate system, the axis that goes "up and down" is called the _____ or the _____.

4. When each number in a set of data is matched with another number by some rule of association, we call it _____.

5. Together, the x-axis and the y-axis form a rectangular _____.

6. The x-axis and the y-axis divide the coordinate system into four regions called _____.

7. The axis lines in a coordinate system intersect at the _____.

8. _____ are the numbers in the ordered pair that specify the location of a point on a rectangular coordinate system.

Name: Date:
Instructor: Section:

Objective 1 Plot a point, given the coordinates, and find the coordinates, given a point.

Video Examples

Review these examples for Objective 1:
1. Use the grid below to plot each point.

   a. (2, 5)

   Start at (0, 0). Move to the right along the horizontal axis until you reach 2. Then move up 5 units so that you are aligned with 5 on the vertical axis. Make a dot and label it (2, 5).

   b. (−5, 4)

   Start at (0, 0). Move to the left along the horizontal axis until you reach −5. Then move up 4 units so that you are aligned with 4 on the vertical axis. Make a dot and label it (−5, 4).

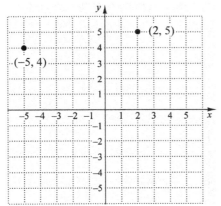

2. Use the grid below to plot each point.

   a. (1, −2)

   Move right 1 unit. Move down 2 units. Make a dot and label it (1, −2).

   b. (−4, 0)

   Move left 4 units. Make a dot and label it (−4, 0).

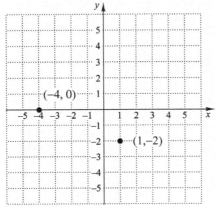

Now Try:
1. Use the grid to plot each point.

   a. (1, 3)

   b. (4, −1)

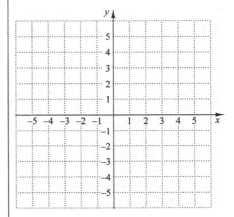

2. Use the grid to plot each point.

   a. (−2, 0)

   b. (0, 5)

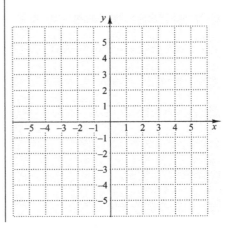

Copyright © 2018 Pearson Education, Inc.

Name: Date:
Instructor: Section:

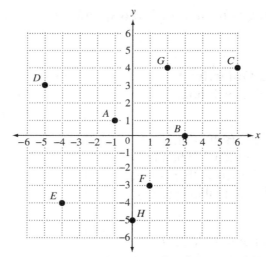

3. Give the coordinates of each point.

a. A

To reach point A from the origin, move 1 unit to the left; then move up 1 unit. The coordinates are (−1, 1).

b. B

To reach point B from the origin, move 3 units to the left; do not move up or down. The coordinates are (3, 0).

c. C

To reach point C from the origin, move 6 units to the right; then move up 4 units. The coordinates are (6, 4).

d. D

To reach point D from the origin, move 5 units to the left; then move up 3 units. The coordinates are (−5, 3).

3. Give the coordinates of each point.

a. E

_____

b. F

_____

c. G

_____

d. H

_____

Name: Date:
Instructor: Section:

Objective 1 Practice Exercises

For extra help, see Examples 1–3 on pages 662–664 of your text.

1. *Plot each point on the rectangular coordinate system below. Label each point with its coordinates.*

   a. (−3, 4)

   b. (2, −1)

   c. (0, −5)

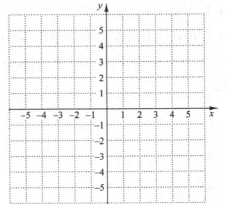

2. *Give the coordinates of each point.*

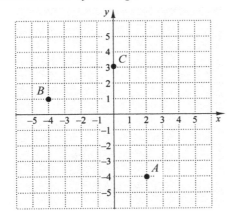

A _____

B _____

C _____

Objective 2 Identify the four quadrants and determine which points lie within each one.

Video Examples

Review these examples for Objective 2:
4. Identify the quadrant in which each point is located.

   a. (5, 11)

   For (5, 11), the pattern is (+, +), so the point is in Quadrant I.

   b. (3, −2)

   For (3, −2), the pattern is (+, −), so the point is in Quadrant IV.

Now Try:
4. Identify the quadrant in which each point is located.

   a. (4, 9)

   _____

   b. (−6, 3)

   _____

Copyright © 2018 Pearson Education, Inc.   289

Name: Date:
Instructor: Section:

c. (−8, 0)

The point corresponding to (−8, 0) is on the *x*-axis, so it isn't in any quadrant.

c. (0, 9)

_____

Objective 2 Practice Exercises

For extra help, see Example 4 on page 664 of your text.

Identify the quadrant in which each point is located.

   3.   (3, −9)   **3.** _____

   4.   (−2, −2)   **4.** _____

   5.   (−3, 0)   **5.** _____

Name: Date:
Instructor: Section:

Chapter 9 GRAPHS AND GRAPHING

9.5 Introduction to Graphing Linear Equations

Learning Objectives
1. Graph linear equations in two variables.
2. Identify the slope of a line as positive or negative.

Key Terms

Use the vocabulary terms listed below to complete each statement in exercises 1–2.

graph a linear equation   **slope**

1. To _____, find at least three ordered pairs that satisfy the equation, and then, plot the points and connect them with a straight line.

2. As you move from left to right, if a line tilts downward, it has a negative _____.

Objective 1   Graph linear equations in two variables.

Video Examples

Review these examples for Objective 1:
1. Graph $x + y = 7$ by finding three solutions and plotting the ordered pairs. Then use the graph to find a fourth solution of the equation.

   Set up a table to organize the information.

x	y	$x + y = 7$	(x, y)
2	5	$2 + 5 = 7$	(2, 5)
3	4	$3 + 4 = 7$	(3, 4)
4	3	$4 + 3 = 7$	(4, 3)

   Plot the ordered pairs and draw a line through the points, extending it in both directions.

   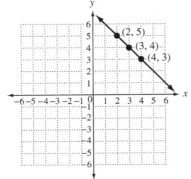

   Find another point. It appears that a point with

Now Try:
1. Graph $x - y = 0$ by finding three solutions and plotting the ordered pairs. Then use the graph to find a fourth solution of the equation.

   _____

   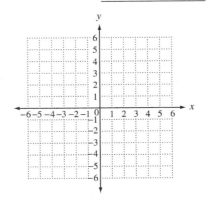

coordinates (5, 2) is a solution.
To check that (5, 2) is a solution, substitute 5 for x and 2 for y in the original equation.

$x + y = 7$

$5 + 2 = 7$

$7 = 7$

The equation balances, so (5, 2) is another solution of $x + y = 7$.

2. Graph $y = 3x$ by finding three solutions and plotting the ordered pairs. Then use the graph to find a fourth solution of the equation.

Set up a table to organize the information.

x	y	$y = 3x$	(x, y)
-1	-3	$-3 = 3(-1)$	$(-1, -3)$
0	0	$0 = 3(0)$	$(0, 0)$
1	3	$3 = 3(1)$	$(1, 3)$

Plot the ordered pairs and draw a line through the points, extending it in both directions.

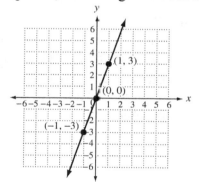

Find another point. It appears that a point with coordinates (2, 6) is a solution.
To check that (2, 6) is a solution, substitute 2 for x and 6 for y in the original equation.

$y = 3x$

$6 = 3(2)$

$6 = 6$

The equation balances, so (2, 6) is another solution of $y = 3x$.

2. Graph $y = 4x - 1$ by finding three solutions and plotting the ordered pairs. Then use the graph to find a fourth solution of the equation.

_____

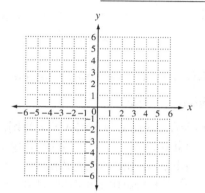

4. Graph $y = -3x + 2$ by finding three solutions and plotting the ordered pairs. Then use the graph to find a fourth solution of the equation.

Set up a table to organize the information.

x	y	$y = -3x + 2$	(x, y)
0	2	$2 = -3(0) + 2$	$(0, 2)$
1	-1	$-1 = -3(1) + 2$	$(1, -1)$
2	-4	$-4 = -3(2) + 2$	$(2, -4)$

Plot the ordered pairs and draw a line through the points, extending it in both directions.

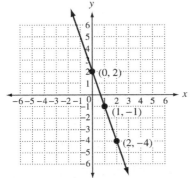

Find another point. It appears that a point with coordinates $(-1, 5)$ is a solution.

To check that $(-1, 5)$ is a solution, substitute -1 for x and 5 for y in the original equation.

$y = -3x + 2$

$5 = -3(-1) + 2$

$5 = 5$

The equation balances, so $(-1, 5)$ is another solution of $y = -3x + 2$.

4. Graph $y = x + 2$ by finding three solutions and plotting the ordered pairs. Then use the graph to find a fourth solution of the equation.

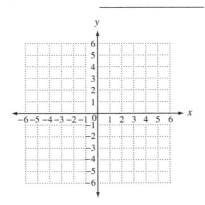

Name: Date:
Instructor: Section:

Objective 1 Practice Exercises

For extra help, see Examples 1–4 on pages 668–672 of your text.

Graph each equation. Make your own table using the listed values of x.

1. $y = x - 4$
   Use 0, 2, and 4 as the values of x.

   1.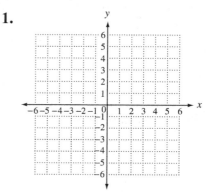

2. $x + y = -3$
   Use –4, –3, and –2 as the values of x.

   2.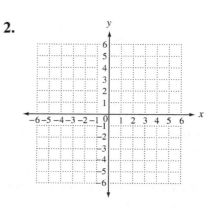

3. $y = \frac{1}{5}x$
   Use –5, 0, and 5 as the values of x.

   3.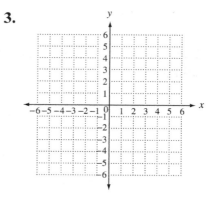

Name: Date:
Instructor: Section:

Objective 2 Identify the slope of a line as positive or negative.

Video Examples

Review this example for Objective 2:
5. Look back at the graph of $y = -3x + 2$ from Example 4. Then complete these sentences.
The graph of $y = -3x + 2$ has a _____ slope.
As the value of x increases, the value of y ____.

The graph of $y = -3x + 2$ has a <u>negative</u> slope (because it tilts downward).
As the value of x increases, the value of y <u>decreases</u> (does the opposite).

Now Try:
5. Look back at the graph of $y = x + 2$ from Example 4. Then complete these sentences. The graph of $y = x + 2$ has a _____ slope. As the value of x increases, the value of y ____.

_____

_____

Objective 2 Practice Exercises

For extra help, see Example 5 on page 674 of your text.

Look back at the graphs in exercises 1–3. Then complete exercises 4–6.

4. The graph of $y = x - 4$ has a ____ slope.   4. _____

5. The graph of $x + y = -3$ has a ____ slope.   5. _____

6. The graph of $y = \frac{1}{5}x$ has a ____ slope.   6. _____

Name: Date:
Instructor: Section:

Chapter 10 EXPONENTS AND POLYNOMIALS

10.1 The Product Rule and Power Rules for Exponents

Learning Objectives
1 Review the use of exponents.
2 Use the product rule for exponents.
3 Use the rule $(a^m)^n = a^{mn}$.
4 Use the rule $(ab)^m = a^m b^m$.
5 Use the rule $\left(\dfrac{a}{b}\right)^m = \dfrac{a^m}{b^m}$.

Key Terms

Use the vocabulary terms listed below to complete each statement in exercises 1–3.

    exponential expression   **base**   **power**

1. 2^5 is read "2 to the fifth _____".

2. A number written with an exponent is called a(n) _____.

3. The _____ is the number being multiplied repeatedly.

Objective 1 Review the use of exponents.

Video Examples

Review these examples for Objective 1:	Now Try:
2. Name the base and exponent of each expression. Then evaluate.	2. Name the base and exponent of each expression. Then evaluate.
a. 3^4	**a.** 2^6

Base: 3
    Exponent: 4
    Value: $3^4 = 3 \cdot 3 \cdot 3 \cdot 3 = 81$
	_____

b. -3^4	**b.** -2^6

Base: 3
    Exponent: 4
    Value: $-3^4 = -1 \cdot (3 \cdot 3 \cdot 3 \cdot 3) = -81$
	_____

c. $(-3)^4$	**c.** $(-2)^6$

Base: -3
    Exponent: 4
    Value: $(-3)^4 = (-3)(-3)(-3)(-3) = 81$
	_____

Name: Date:
Instructor: Section:

Objective 1 Practice Exercises

For extra help, see Examples 1–2 on page 700 of your text.

Write the expression in exponential form and evaluate, if possible.

1. $\left(\frac{1}{3}\right)\left(\frac{1}{3}\right)\left(\frac{1}{3}\right)\left(\frac{1}{3}\right)\left(\frac{1}{3}\right)$

1. _____

Evaluate each exponential expression. Name the base and the exponent.

2. $(-4)^4$

2. _____

base_____

exponent_____

3. -3^8

3. _____

base_____

exponent_____

Objective 2 Use the product rule for exponents.

Video Examples

Review this example for Objective 2:

4. Multiply $5x^4$ and $6x^9$.

$$5x^4 \cdot 6x^9 = (5 \cdot 6) \cdot (x^4 \cdot x^9)$$
$$= 30x^{4+9}$$
$$= 30x^{13}$$

Now Try:

4. Multiply $6x^5$ and $3x^6$.

_____

Name: Date:
Instructor: Section:

Objective 2 Practice Exercises

For extra help, see Examples 3–4 on pages 701–702 of your text.

Use the product rule to simplify each expression, if possible. Write each answer in exponential form.

4. $7^4 \cdot 7^3$ 4. _____

5. $(-2c^7)(-4c^8)$ 5. _____

6. $(3k^7)(-8k^2)(-2k^9)$ 6. _____

Objective 3 Use the rule $(a^m)^n = a^{mn}$.

Video Examples

Review this example for Objective 3:
5d. Use power rule (a) for exponents to simplify.

$(m^3)^4$

$(m^3)^4 = m^{3 \cdot 4}$
$= m^{12}$

Now Try:
5d. Use power rule (a) for exponents to simplify.

$(n^5)^6$

_____

Objective 3 Practice Exercises

For extra help, see Example 5 on page 702 of your text.

Simplify each expression. Write all answers in exponential form.

7. $(7^3)^4$ 7. _____

Name: Date:
Instructor: Section:

8. $-\left(v^4\right)^9$

8. _____

9. $\left[2^3\right]^7$

9. _____

Objective 4   Use the rule $(ab)^m = a^m b^m$.

Video Examples

Review this example for Objective 4:
6b. Use power rule (b) for exponents to simplify.

$$6(rs)^4$$
$$6(rs)^4 = 6\left(r^4 s^4\right)$$
$$= 6r^4 s^4$$

Now Try:
6b. Use power rule (b) for exponents to simplify.
$$7(pq)^3$$

_____

Objective 4 Practice Exercises

For extra help, see Example 6 on page 703 of your text.

Simplify each expression.

10. $\left(5r^3 t^2\right)^4$

10. _____

11. $\left(3a^4 b\right)^3$

11. _____

12. $\left(-2w^3 z^7\right)^4$

12. _____

Name: Date:
Instructor: Section:

Objective 5 Use the rule $\left(\dfrac{a}{b}\right)^m = \dfrac{a^m}{b^m}$.

Video Examples

Review these examples for Objective 5:

7a. Use power rule (c) for exponents to simplify.

$$\left(\dfrac{3}{5}\right)^4$$

$$\left(\dfrac{3}{5}\right)^4 = \dfrac{3^4}{5^4} = \dfrac{81}{625}$$

6c. Use the rules for exponents to simplify.

$$4\left(3m^4 n^5\right)^3$$

$$4\left(3m^4 n^5\right)^3 = 4\left[3^3 \left(m^4\right)^3 \left(n^5\right)^3\right]$$
$$= 4\left[3^3 m^{12} n^{15}\right]$$
$$= 108 m^{12} n^{15}$$

Now Try:

7a. Use power rule (c) for exponents to simplify.

$$\left(\dfrac{5}{6}\right)^3$$

_____

6c. Use the rules for exponents to simplify.

$$8\left(5x^4 y^3\right)^2$$

_____

Objective 5 Practice Exercises

For extra help, see Example 7 on page 704 of your text.

Simplify each expression.

13. $\left(-\dfrac{2x}{5}\right)^3$

13. _____

14. $\left(\dfrac{xy}{z^2}\right)^4$

14. _____

15. $\left(\dfrac{-2a}{b^2}\right)^7$

15. _____

Name: Date:
Instructor: Section:

Chapter 10 EXPONENTS AND POLYNOMIALS

10.2 Integer Exponents and the Quotient Rule

Learning Objectives	
1	Use 0 as an exponent.
2	Use negative numbers as exponents.
3	Use the quotient rule for exponents.
4	Use the product rule with negative exponents.

Key Terms

Use the vocabulary terms listed below to complete each statement in exercises 1–3.

  exponent   **base**   **product rule for exponents**

  power rule for exponents

1. The statement "If m and n are any integers, then $(a^m)^n = a^{mn}$" is an example of the _____.

2. In the expression a^m, a is the _____ and m is the _____.

3. The statement "If m and n are any integers, then $a^m \cdot a^n = a^{m+n}$" is an example of the _____.

Objective 1 Use 0 as an exponent.

Video Examples

Review these examples for Objective 1:	Now Try:
1. Evaluate.	1. Evaluate.
a. $75^0 = 1$	a. 88^0
e. $(-75)^0 = 1$	e. $(-88)^0$
f. $-75^0 = -(1)$ or -1	f. -88^0
c. $9x^0 = 9(1)$, or 9 $(x \neq 0)$	c. $88a^0$ $(a \neq 0)$

Name: Date:
Instructor: Section:

Objective 1 Practice Exercises

For extra help, see Example 1 on page 707 of your text.

Evaluate each expression.

1. -12^0

1. _____

2. $-15^0 - (-15)^0$

2. _____

3. $\dfrac{0^8}{8^0}$

3. _____

Objective 2 Use negative numbers as exponents.

Video Examples

Review these examples for Objective 2:

2. Simplify by writing each expression with positive exponents. Then evaluate the expression, if possible.

   b. 4^{-3}

   $4^{-3} = \dfrac{1}{4^3}$, or $\dfrac{1}{64}$

   c. $5^{-1} - 3^{-1}$

   $5^{-1} - 3^{-1} = \dfrac{1}{5} - \dfrac{1}{3}$

   $= \dfrac{3}{15} - \dfrac{5}{15}$

   $= -\dfrac{2}{15}$

Now Try:

2. Simplify by writing each expression with positive exponents. Then evaluate the expression, if possible.

   b. 3^{-3}

   _____

   c. $4^{-1} - 8^{-1}$

   _____

Objective 2 Practice Exercises

For extra help, see Example 2 on page 708 of your text.

Evaluate or simplify each expression, and write it using only positive exponents. Assume that all variables represent nonzero real numbers.

4. 2^{-4}

4. _____

Name: Date:
Instructor: Section:

5. n^{-9}

5. _____

6. $6^{-1} + 8^{-1}$

6. _____

Objective 3 Use the quotient rule for exponents.

Video Examples

Review these examples for Objective 3:
3. Simplify. Assume that all variables represent nonzero real numbers.

   a. $\dfrac{4^9}{4^6} = 4^{9-6} = 4^3 = 64$

   b. $\dfrac{5^3}{5^7} = 5^{3-7} = 5^{-4} = \dfrac{1}{5^4} = \dfrac{1}{625}$

   d. $\dfrac{p^6}{p^{-4}} = p^{6-(-4)} = p^{10}$

Now Try:
3. Simplify. Assume that all variables represent nonzero real numbers.

   a. $\dfrac{3^{18}}{3^{16}}$

   _____

   b. $\dfrac{2^6}{2^{10}}$

   _____

   d. $\dfrac{z^6}{z^{-4}}$

   _____

Objective 3 Practice Exercises

For extra help, see Example 3 on page 709 of your text.

Use the quotient rule to simplify each expression, and write it using only positive exponents. Assume that all variables represent nonzero real numbers.

7. $\dfrac{13^4}{13^{-3}}$

7. _____

Name: Date:
Instructor: Section:

8. $\dfrac{m^{-3}}{m^{-8}}$

8. _____

9. $\dfrac{z^{-17}}{z^{19}}$

9. _____

Objective 4 Use the product rule with negative exponents.
Video Examples

Review these examples for Objective 4:
4. Simplify each expression. Assume that all variables represent nonzero real numbers. Write answers with positive exponents.

  a. $10^6 \left(10^{-5}\right)$

  $10^6 \left(10^{-5}\right) = 10^{6+(-5)} = 10^1$ or 10

  b. $\left(10^{-7}\right)\left(10^{-4}\right)$

  $\left(10^{-7}\right)\left(10^{-4}\right) = 10^{-7+(-4)} = 10^{-11} = \dfrac{1}{10^{11}}$

  c. $y^{-3} \cdot y^6 \cdot y^{-9}$

  $y^{-3} \cdot y^6 \cdot y^{-9} = y^{-3+6} \cdot y^{-9}$
  $= y^3 \cdot y^{-9}$
  $= y^{3+(-9)}$
  $= y^{-6}$
  $= \dfrac{1}{y^6}$

Now Try:
4. Simplify each expression. Assume that all variables represent nonzero real numbers. Write answers with positive exponents.

  a. $10^7 \left(10^{-6}\right)$

  _____

  b. $\left(10^{-11}\right)\left(10^{-3}\right)$

  _____

  c. $y^{-6} \cdot y^5 \cdot y^{-7}$

  _____

Name: Date:
Instructor: Section:

Objective 4 Practice Exercises

For extra help, see Example 4 on page 710 of your text.

Simplify each expression, and write it using only positive exponents. Assume that all variables represent nonzero real numbers.

10. $3^{-9}(3^4)$

10. _____

11. $p^{-7} \cdot p^4 \cdot p^{15}$

11. _____

12. $x^{-3} \cdot x^2 \cdot x^{-11}$

12. _____

Name: Date:
Instructor: Section:

Chapter 10 EXPONENTS AND POLYNOMIALS

10.3 An Application of Exponents: Scientific Notation

Learning Objectives
1. Express numbers in scientific notation.
2. Convert numbers in scientific notation to numbers without exponents.
3. Use scientific notation in calculations.
4. Solve application problems using scientific notation.

Key Terms

Use the vocabulary terms listed below to complete each statement in exercises 1–3.

    scientific notation   **quotient rule**   **power rule**

1. _____ is used to write very large or very small numbers using exponents.

2. The statement "If m and n are any integers and $b \neq 0$, then $\left(\dfrac{a}{b}\right)^m = \dfrac{a^m}{b^m}$" is an example of the _____.

3. The statement "If m and n are any integers and $b \neq 0$, then $\dfrac{a^m}{a^n} = a^{m-n}$" is an example of the _____.

Objective 1   Express numbers in scientific notation.

Video Examples

Review these examples for Objective 1:	Now Try:
1. Write each number in scientific notation.	1. Write each number in scientific notation.
a. 54,000,000	**a.** 680,000,000
Move the decimal point to follow the 5. Count the number of places the decimal point was moved: 7 places.	
$54,000,000 = 5.4 \times 10^7$	
b. 84,300,000,000	**b.** 47,710,000,000
Move the decimal point 10 places to the left.	
$84,300,000,000 = 8.43 \times 10^{10}$	
c. 7.406	**c.** 9.991
$7.406 = 7.406 \times 10^0$	

Name: Date:
Instructor: Section:

d. 0.00573

The first nonzero digit is 5. Count the places.
Move the decimal point 3 places to the right.
$0.00573 = 5.73 \times 10^{-3}$

d. 0.0463

_____

Objective 1 Practice Exercises

For extra help, see Example 1 on pages 715–716 of your text.

Write each number in scientific notation.

1. 23,651

1. _____

2. −429,600,000,000

2. _____

3. −0.0002208

3. _____

Objective 2   Convert numbers in scientific notation to numbers without exponents.

Video Examples

Review these examples for Objective 2:
2. Write each number without exponents.

   a. 7.4×10^4

   Move the decimal point 4 places to the right, and add three zeros.
   $7.4 \times 10^4 = 74,000$

   c. -8.98×10^{-3}

   Move the decimal point 3 places to the left.
   $-8.98 \times 10^{-3} = -0.00898$

Now Try:
2. Write each number without exponents.

   a. 8.35×10^5

   _____

   c. -1.64×10^{-4}

   _____

Name: Date:
Instructor: Section:

Objective 2 Practice Exercises

For extra help, see Example 2 on page 716 of your text.

Write each number without exponents.

4. -2.45×10^6 4. _____

5. 6.4×10^{-3} 5. _____

6. -4.02×10^0 6. _____

Objective 3   Use scientific notation in calculations.

Video Examples

Review these examples for Objective 3:

3. Perform each calculation. Write answers in scientific notation and also without exponents.

    a. $(8 \times 10^4)(7 \times 10^3)$

$$(8 \times 10^4)(7 \times 10^3) = (8 \times 7)(10^4 \times 10^3)$$
$$= 56 \times 10^7$$
$$= (5.6 \times 10^1) \times 10^7$$
$$= 5.6 \times 10^8$$
$$= 560,000,000$$

    b. $\dfrac{6 \times 10^{-4}}{3 \times 10^2}$

$$\dfrac{6 \times 10^{-4}}{3 \times 10^2} = \dfrac{6}{3} \times \dfrac{10^{-4}}{10^2}$$
$$= 2 \times 10^{-6}$$
$$= 0.000002$$

Now Try:

3. Perform each calculation. Write answers in scientific notation and also without exponents.

    a. $(9 \times 10^5)(3 \times 10^2)$

        _____

    b. $\dfrac{39 \times 10^{-3}}{13 \times 10^5}$

        _____

Name: Date:
Instructor: Section:

Objective 3 Practice Exercises

For extra help, see Example 3 on page 716 of your text.

Perform the indicated operations, and write the answers in scientific notation and also without exponents.

7. $(2.3 \times 10^4) \times (1.1 \times 10^{-2})$

7. _____

8. $\dfrac{9.39 \times 10^1}{3 \times 10^3}$

8. _____

Objective 4 Solve application problems using scientific notation.

Video Examples

Review this example for Objective 4:

5. In 2007, the Gross Domestic Product of the U.S. was 1.38×10^{-3} dollars. That year the U.S. population was about 3.02×10^8 people. What was the per capita GDP, the Gross Domestic Product per person? Round to the nearest hundred.

   Divide the GDP by the population.
   $$\dfrac{1.38 \times 10^{13}}{3.02 \times 10^8} = \dfrac{1.38}{3.02} \times \dfrac{10^{13}}{10^8}$$
   $$= 0.45695 \times 10^5$$
   $$= 4.5695 \times 10^4 \text{ or } 45{,}695$$
   Rounded to the nearest hundred, the GDP is $45,700 per person.

Now Try:

5. Earth has a mass of 6×10^{24} kilograms and a volume of 1.1×10^{21} cubic meters. What is Earth's density in kilograms per cubic meter? Round to the nearest hundredth.

_____

Name: Date:
Instructor: Section:

Objective 4 Practice Exercises

For extra help, see Examples 4–5 on pages 717–718 of your text.

Work each problem. Give answers both in scientific notation and without exponents.

9. Americans eat 6.1 million pretzels in a day. How many pretzels to Americans eat in 1 year?

9. _____

10. An electronic oscillator is producing a signal with a frequency of 1.7×10^8 cycles per second. This frequency is to be tripled. What will the new frequency be?

10. _____

11. The moon has a mass of 7.35×10^{22} kilograms and a volume of 2.2×10^{10} cubic meters. What is the moon's density in kilograms per cubic meter? Round to the nearest hundredth.

11. _____

Name: Date:
Instructor: Section:

Chapter 10 EXPONENTS AND POLYNOMIALS

10.4 Adding and Subtracting Polynomials

Learning Objectives
1. Review combining like terms.
2. Use the vocabulary for polynomials.
3. Evaluate polynomials.
4. Add polynomials.
5. Subtract polynomials.

Key Terms

Use the vocabulary terms listed below to complete each statement in exercises 1–7.

polynomial   descending powers   degree of a term

degree of a polynomial   monomial

binomial   trinomial

1. The _____ is the sum of the exponents on the variables in that term.

2. A polynomial in x is written in _____ if the exponents on x decrease from left to right.

3. A polynomial with exactly three terms is called a _____.

4. A _____ is a term, or the sum of terms, with whole number exponents.

5. A polynomial with exactly one term is called a _____.

6. The _____ is the highest degree of any term of the polynomial.

7. A _____ is a polynomial with exactly two terms.

Objective 1 Review combining like terms.

Video Examples

Review these examples for Objective 1:
1. Simplify each expression by combining like terms.

   a. $-7x^2 + 10x^2$

   $-7x^2 + 10x^2 = (-7 + 10)x^2$
   $= 3x^2$

Now Try:
1. Simplify each expression by combining like terms.

   a. $-6x^4 + 11x^4$

   _____

312   Copyright © 2018 Pearson Education, Inc.

Name: Date:
Instructor: Section:

b. $4x^5 - 15x^5 + x^5$

$4x^5 - 15x^5 + x^5 = (4 - 15 + 1)x^5$
$= -10x^5$

c. $19m^3 + 6m + 5m^3$

$19m^3 + 6m + 5m^3 = (19 + 5)m^3 + 6m$
$= 24m^3 + 6m$

d. $5rs^3 + rs^3 - 4rs^3$

$5rs^3 + rs^3 - 4rs^3 = (5 + 1 - 4)rs^3$
$= 2rs^3$

b. $9x^7 - 18x^7 + x^7$

_____

c. $22m^2 + 15m^3 + 7m^2$

_____

d. $6p^2q - 3p^2q + 2p^2q$

_____

Objective 1 Practice Exercises

For extra help, see Example 1 on page 721 of your text.

In each polynomial, combine like terms whenever possible. Write the result with descending powers.

1. $7z^3 - 4z^3 + 5z^3 - 11z^3$

   1. _____

2. $-1.3z^7 + 0.4z^7 + 2.6z^8$

   2. _____

3. $6c^3 - 9c^2 - 2c^2 + 14 + 3c^2 - 6c - 8 + 2c^3$

   3. _____

Copyright © 2018 Pearson Education, Inc.

Name: Date:
Instructor: Section:

Objective 2 Use the vocabulary for polynomials.

Video Examples

Review these examples for Objective 2:

2. Simplify each polynomial if possible. Then give the degree and tell whether the polynomial is a monomial, a binomial, a trinomial, or none of these.

   a. $5x^4 + 7x$

   We cannot simplify further. This is a binomial of degree 4.

   b. $9x - 7x + 3x$

   $9x - 7x + 3x = 5x$
   The degree is 1. The simplified polynomial is a monomial.

Now Try:

2. Simplify each polynomial if possible. Then give the degree and tell whether the polynomial is a monomial, a binomial, a trinomial, or none of these.

   a. $8x^3 + 4x^2 + 6$

   _____

   b. $x^5 + 3x^5$

   _____

Objective 2 Practice Exercises

For extra help, see Example 2 on page 722 of your text.

For each polynomial, first simplify, if possible, and write the resulting polynomial in descending powers of the variable. Then give the degree of this polynomial, and tell whether it is a monomial, a binomial, a trinomial, or none of these.

4. $3n^8 - n^2 - 2n^8$

   4. _____

   degree: _____

   type: _____

5. $-d^2 + 3.2d^3 - 5.7d^8 - 1.1d^5$

   5. _____

   degree: _____

   type: _____

6. $-6c^4 - 6c^2 + 9c^4 - 4c^2 + 5c^5$

   6. _____

   degree: _____

   type: _____

Name: Date:
Instructor: Section:

Objective 3 Evaluate polynomials.

Video Examples

Review this example for Objective 3:

3. Find the value of $4x^3 + 6x^2 - 5x - 5$ when $x = -3$ and $x = 2$.

    For $x = -3$, replace x with -3.
    $$4x^3 + 6x^2 - 5x - 5$$
    $$= 4(-3)^3 + 6(-3)^2 - 5(-3) - 5$$
    $$= 4(-27) + 6(9) - 5(-3) - 5$$
    $$= -108 + 54 + 15 - 5$$
    $$= -44$$

    Now replace x with 2.
    $$4x^3 + 6x^2 - 5x - 5$$
    $$= 4(2)^3 + 6(2)^2 - 5(2) - 5$$
    $$= 4(8) + 6(4) - 5(2) - 5$$
    $$= 32 + 24 - 10 - 5$$
    $$= 41$$

Now Try:

3. Find the value of $5x^4 + 3x^2 - 9x - 7$ when $x = 4$ and $x = -4$.

    _____

    _____

Objective 3 Practice Exercises

For extra help, see Example 3 on page 722–723 of your text.

Find the value of each polynomial (a) *when x = –2 and* (b) *when x = 3.*

7. $3x^3 + 4x - 19$

    7. a. _____

    b. _____

8. $-4x^3 + 10x^2 - 1$

    8. a. _____

    b. _____

9. $x^4 - 3x^2 - 8x + 9$

    9. a. _____

    b. _____

Name: Date:
Instructor: Section:

Objective 4   Add polynomials.

Video Examples

Review these examples for Objective 4:

4b. Add vertically.

$3x^3 + 7x + 5$ and $x^4 - 6x$

Write like terms and add column by column.

$$\begin{array}{r} 3x^3 + 7x + 5 \\ x^4 - 6x \\ \hline x^4 + 3x^3 + x + 5 \end{array}$$

5a. Find each the by adding horizontally.

Add $5x^4 - 7x^3 + 9$ and $-3x^4 + 8x^3 - 7$

$(5x^4 - 7x^3 + 9) + (-3x^4 + 8x^3 - 7)$
$= 5x^4 - 3x^4 - 7x^3 + 8x^3 + 9 - 7$
$= 2x^4 + x^3 - 2$

Now Try:

4b. Add vertically.

$9x^4 + 3x - 6$ and $7x^2 + 4x$

_____

5a. Find the sum by adding horizontally.

Add $15x^3 - 5x + 3$ and $-11x^3 + 6x + 9$

_____

Objective 4 Practice Exercises

For extra help, see Examples 4–5 on page 723–724 of your text.

Add.

10. $9m^3 + 4m^2 - 2m + 3$
     $\underline{-4m^3 - 6m^2 - 2m + 1}$

10. _____

11. $(x^2 + 6x - 8) + (3x^2 - 10)$

11. _____

12. $(3r^3 + 5r^2 - 6) + (2r^2 - 5r + 4)$

12. _____

Name: Date:
Instructor: Section:

Objective 5 Subtract polynomials.

Video Examples

Review this example for Objective 5:
6b. Perform the subtraction.

Subtract $8x^3 - 5x^2 + 8$ from $9x^3 + 6x^2 - 7$.

$(9x^3 + 6x^2 - 7) - (8x^3 - 5x^2 + 8)$
$= (9x^3 + 6x^2 - 7) + (-8x^3 + 5x^2 - 8)$
$= x^3 + 11x^2 - 15$

Now Try:
6b. Perform the subtraction.

$(7x^3 - 3x - 5) - (18x^3 + 4x - 6)$

_____

Objective 5 Practice Exercises

For extra help, see Example 6 on page 724 of your text.

Subtract.

13. $(-8w^3 + 11w^2 - 12) - (-10w^2 + 3)$ 13. _____

14. $(8b^4 - 4b^3 + 7) - (2b^2 + b + 9)$ 14. _____

15. $(9x^3 + 7x^2 - 6x + 3) - (6x^3 - 6x + 1)$ 15. _____

Name: Date:
Instructor: Section:

Chapter 10 EXPONENTS AND POLYNOMIALS

10.5 Multiplying Polynomials: An Introduction

Learning Objectives
1. Multiply a monomial and a polynomial.
2. Multiply two polynomials.

Key Terms
Use the vocabulary terms listed below to complete each statement in exercises 1–5.

monomial binomial trinomial

polynomial distributive property

1. A _____ is a polynomial with exactly one term.

2. $a(b+c) = ab + ac$ is the statement of the _____.

3. A _____ is a polynomial with exactly three terms.

4. An algebraic expression made up of a term, or the sum of terms, with whole number exponents is called a _____.

5. A polynomial with exactly two terms is a _____.

Objective 1 Multiply a monomial and a polynomial.

Video Examples

Review these examples for Objective 1:	Now Try:
1. Find each product.	1. Find each product.
a. $5x^2(7x+3)$	**a.** $8x^3(4x+8)$
Use the distributive property.	
$5x^2(7x+3) = 5x^2(7x) + 5x^2(3)$	
$\qquad = 35x^3 + 15x^2$	
b. $-9n^4(6n^4 + 5n^3 + 7n - 1)$	**b.** $-7m^5(5m^3 - 6m^2 + 4m - 1)$
Use the distributive property.	
$-9n^4(6n^4 + 5n^3 + 7n - 1)$	
$= -9n^4(6n^4) - 9n^4(5n^3) - 9n^4(7n) - 9n^4(-1)$	
$= -54n^8 - 45n^7 - 63n^5 + 9n^4$	

Name: Date:
Instructor: Section:

Objective 1 Practice Exercises

For extra help, see Example 1 on page 729 of your text.

Find each product.

1. $7z(5z^3 + 2)$ 1. _____

2. $2m(3 + 7m^2 + 3m^3)$ 2. _____

3. $-3y^2(2y^3 + 3y^2 - 4y + 11)$ 3. _____

Objective 2   Multiply two polynomials.

Video Examples

Review these examples for Objective 2:

2. Find each product.

   a. Multiply $(x+7)(x-5)$.

   Multiply each term of the second polynomial by each terms of the first. Then combine like terms.
   $(x+7)(x-5) = x(x) + x(-5) + 7(x) + 7(-5)$
   $= x^2 + (-5x) + 7x + (-35)$
   $= x^2 + 2x - 35$

   b. Multiply $(x^2 + 6)(5x^3 - 4x^2 + 3x)$.

   Multiply each term of the second polynomial by each term of the first.
   $(x^2 + 6)(5x^3 - 4x^2 + 3x)$
   $= x^2(5x^3) + x^2(-4x^2) + x^2(3x)$
   $\quad + 6(5x^3) + 6(-4x^2) + 6(3x)$
   $= 5x^5 - 4x^4 + 3x^3 + 30x^3 - 24x^2 + 18x$
   $= 5x^5 - 4x^4 + 33x^3 - 24x^2 + 18x$

Now Try:

2. Find each product.

   a. Multiply $(x+9)(x-6)$.

   _____

   b. Multiply $(x^3 + 9)(4x^4 - 2x^2 + x)$.

   _____

Name: Date:
Instructor: Section:

3. Multiply $(2x^3 + 7x^2 + 5x - 1)(4x + 6)$ using the vertical method.

   Write the polynomials vertically.
   $$\begin{array}{r} 2x^3 + 7x^2 + 5x - 1 \\ 4x + 6 \\ \hline \end{array}$$

   Begin by multiplying each term in the top row by 6.
   $$\begin{array}{r} 2x^3 + 7x^2 + 5x - 1 \\ 4x + 6 \\ \hline 12x^3 + 42x^2 + 30x - 6 \end{array}$$

   Now multiply each term in the top row by $4x$. Then add like terms.
   $$\begin{array}{r} 2x^3 + 7x^2 + 5x - 1 \\ 4x + 6 \\ \hline 12x^3 + 42x^2 + 30x - 6 \\ 8x^4 + 28x^3 + 20x^2 - 4x \\ \hline 8x^4 + 40x^3 + 62x^2 + 26x - 6 \end{array}$$

   The product is $8x^4 + 40x^3 + 62x^2 + 26x - 6$.

3. Multiply $(4x^3 - 3x^2 + 6x + 5)(7x - 3)$ using the vertical method.

   3. _____

Objective 2 Practice Exercises

For extra help, see Examples 2–3 on page 730 of your text.

Find each product.

4. $(x + 3)(x^2 - 3x + 9)$ 4. _____

5. $(2m^2 + 1)(3m^3 + 2m^2 - 4m)$ 5. _____

6. $(x - 5)(x + 4)$ 6. _____

Chapter 1 INTRODUCTION TO ALGEBRA: INTEGERS

1.1 Place Value

Key Terms
1. digits
2. whole numbers
3. place value system

Objective 1
  Now Try
  1. 0, 5, 59, 350

  Practice Exercises
  1. 2
  3. 1,365

Objective 2
  Now Try
  2. hundred-thousands and hundreds

  Practice Exercises
  5. ten-millions

Objective 3
  Now Try
3a. nine million, seventy-five thousand, eight hundred sixty-two

3b. fifty-four trillion, eight hundred billion, five hundred forty-three million, seven hundred thousand, one hundred

4a. 753,006
4b. 11,000,010,025

  Practice Exercises
  7. fifty-nine billion, five hundred four million, eight hundred six thousand, eight hundred seventy-three

  9. 987,000,330

1.2 Introduction to Integers

Key Terms
1. absolute value
2. number line
3. integers

Objective 1
  Now Try
1a. −150 feet
1b. +68°F or 68°F

  Practice Exercises
  1. −46
  3. −1,200,000

Answers

Objective 2
 Now Try
 2. [number line with points at -5, -4, -2, 0, 3, 4 on scale -5 to 5]

 Practice Exercises
 5. [number line with points at -4, -2, -1, 1, 2, 3 on scale -5 to 5]

Objective 3
 Now Try
 3a. $0 > -8$ 3b. $-3 > -9$ 3c. $11 > -4$

 Practice Exercises
 7. $>$ 9. $<$

Objective 4
 Now Try
 4a. 12   4b. 13   4c. 0

 Practice Exercises
 11. 21

1.3 Adding Integers

Key Terms
 1. commutative property of addition
 2. addends
 3. associative property of addition
 4. sum
 5. addition property of 0

Objective 1
 Now Try
 1. -5 2a. -36 2b. 40
 3a. -16 3b. 21

 Practice Exercises
 1. -3 3. 18 points

Objective 2
 Now Try
 6a. $-20 + (19 + (-19)) = -20 + 0 = -20$
 6b. $8 + 2 + (-17) = (8 + 2) + (-17) = 10 + (-17) = -7$

 Practice Exercises
 5. $8 + (-15 + (-5)) = -12$

322   Copyright © 2018 Pearson Education, Inc.

Answers

1.4 Subtracting Integers

Key Terms
1. opposite
2. additive inverse

Objective 1
  Now Try
1a. -25, $25+(-25)=0$

  Practice Exercises
1. 11; $-11+11=0$
3. 5; $-5+5=0$

Objective 2
  Now Try
2a. $13+(-15)=-2$
2d. $-8+(-16)=-24$
2b. $-10+9=-1$
2c. $12+11=23$

  Practice Exercises
5. -14

Objective 3
  Now Try
3. -38

  Practice Exercises
7. 0
9. -22

1.5 Problem Solving: Rounding and Estimating

Key Terms
1. front end rounding
2. rounding
3. estimate

Objective 1
  Now Try
1b. $\underline{6}42$; 600

  Practice Exercises
1. 3
3. 4

Objective 2
  Now Try
4a. -4740
5a. $-80,000$

  Practice Exercises
5. 810,000

Objective 3
  Now Try
6b. 1,000,000
7. $2200; $2412

  Practice Exercises
7. 200,000
9. 30°; 35°

Answers

1.6 Multiplying Integers

Key Terms
1. commutative property of multiplication
2. factors
3. associative property of multiplication
4. multiplication property of 0
5. product
6. distributive property
7. multiplication property of 1

Objective 1
  Now Try
1b. $8 \cdot 15$ or $8(15)$ or $(8)(15)$. The factors are 8 and 15. The product is 120.

  Practice Exercises
1. $-3 \cdot 6$; $(-3)(6)$ 3. $4 \cdot 5$; $(4)(5)$

Objective 2
  Now Try
2a. -42 2b. 60 3b. -64
2c. -54

  Practice Exercises
5. -62

Objective 3
  Now Try
4b. -92; multiplication property of 1

5a. $63 = 63$; commutative property of multiplication

5b. $-198 = -198$; associative property of multiplication

6b. $-9(-5) + (-9)(3)$; both results are 18

  Practice Exercises
7. 0   9. $3 \cdot ((-2) \cdot (-9)) = 54$

Objective 4
  Now Try
7. -1800; -1785

  Practice Exercises
11. $1300; $1315

Answers

1.7   Dividing Integers

Key Terms
1. factors   2. quotient   3. product

Objective 1
  Now Try
1a. −5   1b. 4   1c. −3

  Practice Exercises
1. −7   3. −10

Objective 2
  Now Try
2a. 1 Any nonzero number divided by itself is 1.

2b. 93 Any number divided by 1 is the number.

2c. 0 Zero divided by any nonzero number is 0.

2d. undefined Division by 0 is undefined.

  Practice Exercises
5. 1. Any nonzero number divided by itself is 1.

Objective 3
  Now Try
3c. −1   3a. 10

  Practice Exercises
7. 2   9. −21

Objective 4
  Now Try
4. Est: $400; Exact: $309

  Practice Exercises
11. $500; $625

Objective 5
  Now Try
5b. 9 trips, 8 elevator trips would leave 6 people walking to the top

  Practice Exercises
13. $96 \div 7 = 13$ R5. Every truck has at least 13 construction workers. The remainder of 5 means that 5 trucks each carry a 14th worker.

15. $1500 \div 468 = 3$ R 96. This means that each student receives three tickets, and there are 96 seats that are available for assignment.

Answers

1.8 Exponents and Order of Operations

Key Terms
1. order of operations   2. exponent

Objective 1
  Now Try
1a. 5^4; 625; 5 to the fourth power

1b. 6^2; 36; 6 squared, or 6 to the second power

1c. 23^1; 23; 23 to the first power

  Practice Exercises
  1. 12^3; twelve cubed   3. 7^7; seven to the seventh power

Objective 2
  Now Try
2a. 49   2c. 81   2b. −1000
  Practice Exercises
  5. 144

Objective 3
  Now Try
6a. −17   4. 13   6b. −53
5a. −24
  Practice Exercises
  7. 17   9. −16

Objective 4
  Now Try
  7. −9
  Practice Exercises
  11. 16

Answers

Chapter 2 UNDERSTANDING VARIABLES AND SOLVING EQUATIONS

2.1 Introduction to Variables

Key Terms
1. expression
2. variable
3. evaluate the expression
4. constant
5. coefficient

Objective 1
Now Try
1. Variable: c; constant: 5; $c + 5$

Practice Exercises
1. -7: constant; h: variable
3. 9: coefficient; k: variable; 1: constant

Objective 2
Now Try
2a. 29; Order 29 lunches.
2b. 50; Order 50 lunches
3. 32 yards
4. 128
5a. 96

Practice Exercises
5. (a) $10,040 (b) $12,620

Objective 3
Now Try
6. $\frac{0}{b} = 0$

Practice Exercises
7. Multiplication is distributive over addition.
9. Zero added to any number equals that number.

Objective 4
Now Try
7a. $z \cdot z \cdot z \cdot z$
7b. $14 \cdot r \cdot s \cdot s$
7c. $-13 \cdot m \cdot m \cdot m \cdot n \cdot n \cdot n \cdot n \cdot n$
8b. 144
8c. -240

Practice Exercises
11. $u \cdot u \cdot u \cdot v \cdot v \cdot w \cdot w$

Answers

2.2 Simplifying Expressions

Key Terms
1. like terms
2. term
3. simplify an expression
4. variable term
5. constant term

Objective 1
Now Try
2a. $20x$

Practice Exercises
1. like terms: $2x, -3x$; coefficients: $2, -3$
3. $-19c^3z^4$

Objective 2
Now Try
3b. $7c - 9$
4c. $40y^2$

Practice Exercises
5. $-6s - 5st - 16t + 8$

Objective 3
Now Try
5c. $28x - 112$
5a. $30x + 12$
5b. $-54x - 36$
6. $6x - 6$

Practice Exercises
7. $-5t - 20$
9. $20q - 3$

2.3 Solving Equations Using Addition

Key Terms
1. solution
2. addition property of equality
3. equation
4. solve an equation
5. check the solution

Objective 1
Now Try
1. 29

Practice Exercises
1. 12
3. 7

Objective 2
Now Try
2b. $x = -2$
2a. $c = 16$

Practice Exercises
5. $n = -66$

Answers

Objective 3
  Now Try
  3a. $x = -11$ 3b. $b = 6$
  Practice Exercises
  7. $a = 3$ 9. $n = -44$

2.4 Solving Equations Using Division

Key Terms
  1. division property of equality 2. addition property of equality

Objective 1
  Now Try
  1b. $w = -16$
  Practice Exercises
  1. $m = -7$ 3. $h = 28$

Objective 2
  Now Try
  2a. $y = 4$ 2b. $h = 0$
  Practice Exercises
  5. $m = 33$

Objective 3
  Now Try
  3. $y = -13$
  Practice Exercises
  7. $b = -19$ 9. $v = -4$

2.5 Solving Equations with Several Steps

Key Terms
  1. addition property of equality 2. distributive property
  3. division property of equality

Objective 1
  Now Try
  1. $n = 7$ 2. $x = -10$
  Practice Exercises
  1. $m = 0$ 3. $q = -3$

Objective 2
  Now Try
  3. $m = 0$ 4. $m = -5$
  Practice Exercises
  5. $t = 6$

Answers

Chapter 3 SOLVING APPLICATION PROBLEMS

3.1 Problem Solving: Perimeter

Key Terms
1. triangle
2. perimeter
3. formula
4. square
5. rectangle
6. parallelogram

Objective 1
  Now Try
  1. 60 in.
  2. 125 cm
  Practice Exercises
  1. 28 ft
  3. 19 mi

Objective 2
  Now Try
  3. 184 ft
  4. 18 in.
  Practice Exercises
  5. 13 ft

Objective 3
  Now Try
  5. 74 m
  7. 306 in.
  Practice Exercises
  7. 22 in.
  9. 255 cm

3.2 Problem Solving: Area

Key Terms
1. square
2. parallelogram
3. rectangle
4. area

Objective 1
  Now Try
  1b. 189 cm^2
  2. 5 ft
  Practice Exercises
  1. 10 cm^2
  3. 76 cm

Objective 2
  Now Try
  3. 169 in.^{2}
  Practice Exercises
  5. 2025 cm^2

Answers

Objective 3
Now Try
5a. 171 cm^2 6. 9 ft

Practice Exercises
7. 208 m^2 9. 6 m

Objective 4
Now Try
7. 1688 m; 169,984 m^2

Practice Exercises
11. 8 ft

3.3   Solving Application Problems with One Unknown Quantity

Key Terms
1. sum; increased by   2. product; double
3. quotient; per   4. difference; less than

Objective 1
Now Try
1a. $x + 7$ or $7 + x$ 1b. $43 + x$ or $x + 43$ 1c. $x + 63$ or $63 + x$
1d. $-20 + x$ or $x + (-20)$ 1e. $x + 75$ or $75 + x$ 1f. $x - 44$
1g. $89 - x$ 1h. $x - 89$ 1i. $x - 16$
1j. $x - 36$ 1k. $48 - x$ 2a. $57x$
2b. $43x$ 2c. $2x$ 2d. $\dfrac{-13}{x}$
2e. $\dfrac{x}{19}$ 2f. $6x - 81$

Practice Exercises
1. $-6x$ 3. $1 + 3x$ or $3x + 1$

Objective 2
Now Try
3. $7x + 21 = 56$; $x = 5$

Practice Exercises
5. $3 + 7x = 31$; $x = 4$

Objective 3
Now Try
5. 55 paper plates

Practice Exercises
7. 12 bananas   9. 7 celery sticks

Answers

3.4 Solving Application Problems with Two Unknown Quantities

Key Terms

1. added to; more than
2. times; triple
3. divided by; half
4. subtracted from; minus

Objective 1

Now Try

3. length: 31 in.; width: 12 in.
2. 38 m, 60 m

Practice Exercises

1. Greer: 8135 votes; Jones: 8651 votes
3. length: 36 yd; width: 12 yd

Chapter 4 RATIONAL NUMBERS: POSITIVE AND NEGATIVE FRACTIONS

4.1 Introduction to Signed Fractions

Key Terms
1. equivalent fractions
2. fraction
3. improper fraction
4. numerator
5. proper fraction
6. denominator

Objective 1
 Now Try
 1. shaded: $\frac{3}{7}$; unshaded: $\frac{4}{7}$
 2. $\frac{9}{8}$

 Practice Exercises
 1. $\frac{3}{5}$; $\frac{2}{5}$
 3. $\frac{3}{5}$; $\frac{7}{5}$

Objective 2
 Now Try
3b. Numerator: 6; denominator: 23; 23 equal parts in the whole

4a. $\frac{7}{8}, \frac{9}{11}, \frac{12}{29}, \frac{1}{5}$
4b. $\frac{22}{7}, \frac{13}{13}, \frac{17}{4}$

 Practice Exercises
 5. 12; 5

Objective 3
 Now Try
 5.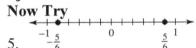

 Practice Exercises
 7.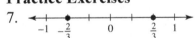
 9.

Objective 4
 Now Try
 6. $\frac{9}{10}, \frac{9}{10}$

 Practice Exercises
 11. $\frac{8}{9}$

Objective 5
 Now Try
 7a. $-\frac{24}{30}$
 7b. $\frac{5}{6}$

 Practice Exercises
 13. $-\frac{32}{48}$
 15. $-\frac{1}{5}$

Answers

4.2 Writing Fractions in Lowest Terms

Key Terms
1. composite number   2. prime factorization   3. prime number
4. lowest terms

Objective 1
Now Try
1a. yes   1b. no; 9

Practice Exercises
1. yes   3. no; 4

Objective 2
Now Try
2a. $\dfrac{5}{9}$ 2b. $\dfrac{7}{8}$ 2c. $-\dfrac{5}{13}$

2d. $\dfrac{11}{12}$

Practice Exercises
5. $\dfrac{5}{7}$

Objective 3
Now Try
3. Prime: 29 and 31; composite: 8 and 21; neither: 1   4a. $2 \cdot 2 \cdot 2 \cdot 2 \cdot 2 \cdot 3$

5a. $2 \cdot 2 \cdot 2 \cdot 19$ 5b. $2 \cdot 2 \cdot 2 \cdot 3 \cdot 7$

Practice Exercises
7. $3 \cdot 5 \cdot 7$ 9. $2 \cdot 2 \cdot 2 \cdot 2 \cdot 2 \cdot 2 \cdot 5$

Objective 4
Now Try
6a. $\dfrac{1}{3}$ 6b. $\dfrac{3}{10}$

Practice Exercises
11. $\dfrac{3 \cdot 3 \cdot 7}{3 \cdot 5 \cdot 7}; \dfrac{3}{5}$

Objective 5
Now Try
7a. $\dfrac{5}{6x}$ 7b. $\dfrac{1}{3}$ 7c. $\dfrac{b^2}{4a}$

Practice Exercises
13. $\dfrac{3r}{s^2}$ 15. $\dfrac{4}{3x}$

Answers

4.3 Multiplying and Dividing Signed Fractions

Key Terms
1. reciprocals
2. indicator words
3. of
4. each

Objective 1
Now Try
1a. $\dfrac{40}{143}$
1b. $-\dfrac{45}{88}$
2b. $\dfrac{3}{20}$

Practice Exercises
1. $-\dfrac{1}{7}$
3. 45

Objective 2
Now Try
4a. $\dfrac{14}{15}$
4b. $\dfrac{2c}{15}$

Practice Exercises
5. $\dfrac{5}{2m}$

Objective 3
Now Try
5a. $\dfrac{1}{12}$

Practice Exercises
7. $-\dfrac{1}{24}$
9. $\dfrac{6}{7}$

Objective 4
Now Try
6b. $\dfrac{10}{9b^2}$
6a. $\dfrac{5a}{b}$

Practice Exercises
11. $6x$

Objective 5
Now Try
7a. $\dfrac{21}{22}$ m^2

Practice Exercises
13. $600
15. 72 servings

Answers

4.4 Adding and Subtracting Signed Fractions
Key Terms
1. unlike fractions
2. like fractions
3. least common denominator

Objective 1
Now Try
1a. $\dfrac{2}{3}$
1b. $\dfrac{1}{2}$
1c. $-\dfrac{3}{5}$
1d. $\dfrac{4}{a^2}$

Practice Exercises
1. $-\dfrac{2}{3}$
3. $\dfrac{4}{y^2}$

Objective 2
Now Try
2a. 27
2b. 30
3a. 100
3b. 90

Practice Exercises
5. 63

Objective 3
Now Try
4c. $\dfrac{41}{72}$
4d. $\dfrac{61}{7}$
4b. $-\dfrac{1}{6}$

Practice Exercises
7. $\dfrac{3}{10}$
9. $\dfrac{8}{15}$

Objective 4
Now Try
5a. $\dfrac{10+3x}{15}$
5b. $\dfrac{5x-64}{8x}$

Practice Exercises
11. $\dfrac{1+2x}{6}$

Answers

4.5 Problem Solving: Mixed Numbers and Estimating

Key Terms
1. mixed number
2. improper fraction

Objective 1
Now Try

1.

Practice Exercises

1.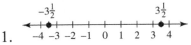

3.

Objective 2
Now Try

2. $\dfrac{35}{6}$

3b. $4\dfrac{1}{9}$

Practice Exercises

5. $-\dfrac{16}{9}$

Objective 3
Now Try

5a. $20;\ 17\dfrac{1}{2}$

6a. $\dfrac{4}{7};\ \dfrac{2}{3}$

Practice Exercises

7. $4;\ 4\dfrac{4}{21}$

9. $2\dfrac{1}{2};\ 2\dfrac{2}{9}$

Objective 4
Now Try

7a. $6;\ 6\dfrac{5}{9}$

7b. $3;\ 3\dfrac{7}{12}$

7c. $5;\ 4\dfrac{6}{7}$

Practice Exercises

11. $4;\ 3\dfrac{5}{9}$

Objective 5
Now Try

8a. 6 oz; $7\dfrac{25}{48}$ oz

Practice Exercises

13. $12\ m^2;\ 12\dfrac{1}{2}\ m^2$

15. 3 in.; $3\dfrac{8}{15}$ in.

Answers

4.6 Exponents, Order of Operations, and Complex Fractions

Key Terms
1. order of operations
2. exponent
3. complex fraction

Objective 1
  Now Try
  1a. $-\dfrac{1}{64}$ 1b. $\dfrac{5}{28}$

  Practice Exercises
  1. $-\dfrac{1}{9}$ 3. $\dfrac{1}{36}$

Objective 2
  Now Try
  2a. $\dfrac{11}{30}$ 2b. $-\dfrac{77}{81}$

  Practice Exercises
  5. $\dfrac{3}{4}$

Objective 3
  Now Try
  3a. $\dfrac{20}{21}$

  Practice Exercises
  7. -8 9. $\dfrac{25}{36}$

4.7 Problem Solving: Equations Containing Fractions

Key Terms
1. division property of equality
2. multiplication property of equality

Objective 1
  Now Try
  1a. 21 1b. -30 1c. $\dfrac{2}{5}$

  Practice Exercises
  1. $b = -36$ 3. $h = \dfrac{3}{16}$

Objective 2
  Now Try
  2a. 30 2b. -21

Answers

Practice Exercises
5. $r = -4$

Objective 3
Now Try
3. 80 in. or 6 ft 8 in.

Practice Exercises
7. 20 years old
9. 30 years old

4.8 Geometry Applications: Area and Volume

Key Terms
1. area
2. volume

Objective 1
Now Try
1a. $\frac{77}{128}$ in.^{2}

Practice Exercises
1. $P = 66$ m; $A = 200$ m^2
3. 1940 yd^2

Objective 2
Now Try
3b. $62\frac{1}{2}$ in.^{3}

Practice Exercises
5. $166\frac{3}{8}$ ft^3

Objective 3
Now Try
4. 25.8 ft^3

Practice Exercises
7. 196 ft^3
9. $139\frac{1}{2}$ m^3

Answers

Chapter 5 RATIONAL NUMBERS: POSITIVE AND NEGATIVE DECIMALS

5.1 Reading and Writing Decimal Numbers

Key Terms
1. decimals
2. place value
3. decimal point

Objective 1
  Practice Exercises
  1. $\frac{8}{10}$; 0.8; eight tenths
  3. $\frac{58}{100}$; 0.58; fifty-eight hundredths

Objective 2
  Now Try
  2a. 8 hundreds; 6 tens; 2 ones; 9 tenths; 3 hundredths
  2b. 0 ones; 0 tenths; 0 hundredths; 7 thousandths; 6 ten-thousandths; 9 hundred-thousandths

  Practice Exercises
  5. 7; 1

Objective 3
  Now Try
  3c. seven hundredths
  4b. eighteen and nine thousandths

  Practice Exercises
  7. eight hundredths
  9. ninety seven and eight thousandths

Objective 4
  Now Try
  6c. $6\frac{1}{25}$
  6a. $\frac{1}{5}$

  Practice Exercises
  11. $3\frac{3}{5}$

340 Copyright © 2018 Pearson Education, Inc.

Answers

5.2 Rounding Decimal Numbers

Key Terms
1. decimal places
2. rounding

Objective 1

Objective 2
Now Try
1. 43.803
2d. 0.6
2a. 0.80

Practice Exercises
1. 489.8
3. 989.990

Objective 3
Now Try
3a. $2.08
3b. $425.10
4c. $881
4e. $1

Practice Exercises
5. $11,840

5.3 Adding and Subtracting Signed Decimal Numbers

Key Terms
1. front end rounding
2. estimating

Objective 1
Now Try
1b. 21.709
2b. 1013.931
4c. 0.501

Practice Exercises
1. 92.49
3. 72.453

Objective 2
Now Try
5b. −10.586
6a. −11.74
5a. −38.4

Practice Exercises
5. 178

Objective 3
Now Try
7a. 14; 13.847
7b. $50; $41.55

Practice Exercises
7. 680; 676.60
9. −15; −14.562

Answers

5.4 Multiplying Signed Decimal Numbers

Key Terms
1. factor
2. decimal places
3. product

Objective 1
 Now Try
 1. −10.793
 2. 0.00186

 Practice Exercises
 1. −90.71
 3. 0.0037

Objective 2
 Now Try
 3. 100; 112.8229

 Practice Exercises
 5. 600; 756.6478

5.5 Dividing Signed Decimal Numbers

Key Terms
1. dividend
2. repeating decimal
3. quotient
4. divisor

Objective 1
 Now Try
 1a. −1.413
 2. 20.178
 3. 16.791

 Practice Exercises
 1. 6.96
 3. 2.359

Objective 2
 Now Try
 4a. 20,410

 Practice Exercises
 5. 128.25

Objective 3
 Now Try
 5. 40; 37.8

 Practice Exercises
 7. reasonable
 9. reasonable

Objective 4
 Now Try
 6b. 22.62
 6c. −0.233

 Practice Exercises
 11. 53.24

Answers

5.6 Fractions and Decimals

Key Terms
1. equivalent
2. numerator
3. denominator
4. mixed number

Objective 1
  Now Try
  1a. 0.0625   1b. 4.375   2. 0.417
  Practice Exercises
  1. 0.875   3. 19.708

Objective 2
  Now Try
  3a. >   3b. =   3c. <
  4a. 0.3, 0.3057, 0.307
  Practice Exercises
  5. $\frac{3}{11}$, 0.29, $\frac{1}{3}$

5.7 Problem Solving with Statistics: Mean, Median, and Mode

Key Terms
1. mode
2. weighted mean
3. mean
4. median

Objective 1
  Now Try
  2. 52.7
  Practice Exercises
  1. 60.3   3. 273.1

Objective 2
  Now Try
  3. 40.2
  Practice Exercises
  5. 4.7

Objective 3
  Now Try
  5. 18   6. 0.06
  Practice Exercises
  7. 232   9. 239.5

Objective 4
  Now Try
  7b. 13 and 16
  Practice Exercises
  11. 24, 35, 39

Answers

5.8   Geometry Applications: Pythagorean Theorem and Square Roots

Key Terms
1. right triangle   2. hypotenuse   3. square root
4. legs

Objective 1
 Now Try
1a. 4.47   1b. 9.59
 Practice Exercises
1. 4.123   3. 10.100

Objective 2
 Now Try
2a. 9.8 in.   2b. 12 km
 Practice Exercises
5. 2.2 cm

Objective 3
 Now Try
3. 9.5 ft
 Practice Exercises
7. 10.3 ft   9. 8 ft

5.9   Problem Solving: Equations Containing Decimals

Key Terms
1. division property of equality
2. addition property of equality

Objective 1
 Now Try
1a. $w = -13.9$ 1b. $x = 19.6$
 Practice Exercises
1. $n = -4.6$ 3. $h = 7.7$

Objective 2
 Now Try
2a. $x = 3.05$ 2b. $t = -8.5$
 Practice Exercises
5. $r = 2.5$

Objective 3
 Now Try
3b. $x = 0.53$
 Practice Exercises
7. $y = -15$ 9. $x = 29.4$

Answers

Objective 4
  Now Try
  4. 32 min
  Practice Exercises
11. 30 mg

5.10 Geometry Applications: Circles, Cylinders, and Surface Area

Key Terms
1. radius   2. circumference   3. circle
4. π (pi)   5. surface area   6. diameter

Objective 1
  Now Try
1b. 29.5 in.
  Practice Exercises
1. 4 ft   3. $6\frac{1}{4}$ yd

Objective 2
  Now Try
2a. 138.2 yd
  Practice Exercises
5. 28.3 yd

Objective 3
  Now Try
3b. 1519.8 yd^2
  Practice Exercises
7. 22.3 yd^2 9. 1061.3 m^2

Objective 4
  Now Try
7a. 1846.3 in^3
  Practice Exercises
11. 6358.5 cm^3

Objective 5
  Now Try
8. $V = 7956$ mm^3; $S = 2670$ mm^2
  Practice Exercises
13. 558 in.^{2} 15. 85.7 in.^{2}

Objective 6
  Now Try
9. $V \approx 113.0$ ft^3; $S \approx 138.2$ ft^2
  Practice Exercises
17. 183.6 ft^2

Answers

Chapter 6 RATIO, PROPORTION, AND LINE/ANGLE/TRIANGLE RELATIONSHIPS

6.1 Ratios

Key Terms

1. ratio
2. numerator; denominator

Objective 1

Now Try

2a. $\dfrac{2}{3}$

Practice Exercises

1. $\dfrac{25}{19}$
3. $\dfrac{1}{4}$

Objective 2

Now Try

4b. $\dfrac{22}{17}$

Practice Exercises

5. $\dfrac{9}{2}$

Objective 3

Now Try

5a. $\dfrac{5}{7}$

Practice Exercises

7. $\dfrac{2}{7}$
9. $\dfrac{5}{6}$

6.2 Rates

Key Terms

1. unit rate
2. cost per unit
3. rate

Objective 1

Now Try

1a. $\dfrac{1 \text{ dollar}}{5 \text{ pages}}$
1b. $\dfrac{15 \text{ strokes}}{1 \text{ minute}}$
1c. $\dfrac{33 \text{ strawberries}}{2 \text{ cakes}}$

Practice Exercises

1. $\dfrac{7 \text{ pills}}{1 \text{ patient}}$
3. $\dfrac{32 \text{ pages}}{1 \text{ chapter}}$

Answers

Objective 2
Now Try
2a. 28 miles/gallon   2c. $145/day

Practice Exercises
5. $225/pound

Objective 3
Now Try
3. 5 pints at $1.70/pint   4a. eight-pack

Practice Exercises
7. 24 ounces for $2.08   9. 5 cans for $2.75

6.3   Proportions
Key Terms
1. proportion   2. cross products   3. ratio

Objective 1
Now Try
1a. $\dfrac{24}{17} = \dfrac{72}{51}$ 1b. $\dfrac{\$10}{7 \text{ cans}} = \dfrac{\$60}{42 \text{ cans}}$

Practice Exercises
1. $\dfrac{50}{8} = \dfrac{75}{12}$ 3. $\dfrac{3}{33} = \dfrac{12}{132}$

Objective 2
Now Try
2a. $\dfrac{9}{7} \neq \dfrac{4}{3}$, false   2b. $\dfrac{1}{3} = \dfrac{1}{3}$, true   3a. $306 = 306$, true

3b. $32 \neq 35$, false

Practice Exercises
5. $\dfrac{6}{5} = \dfrac{6}{5}$; true

Objective 3
Now Try
4a. 32   4b. 10.67   5a. $2\dfrac{3}{5}$

Practice Exercises
7. 36   9. 21

Answers

6.4 Problem Solving with Proportions

Key Terms
1. ratio
2. rate

Objective 1
  Now Try
  1. $\frac{23 \text{ hr}}{4 \text{ apt}} = \frac{x \text{ hr}}{16 \text{ apt}}$; $x = 92$ hr
  2. $\frac{4}{5} = \frac{x}{540}$; $x = 432$ people

  Practice Exercises
  1. $108
  3. approximately 333 deer

6.5 Geometry: Lines and Angles

Key Terms
1. ray
2. perpendicular lines
3. obtuse angle
4. point
5. angle
6. line
7. acute angle
8. degrees
9. parallel lines
10. line segment
11. intersecting lines
12. right angle
13. vertical angles
14. congruent angles
15. supplementary angles
16. complementary angles
17. alternate interior angles
18. corresponding angles
19. straight angle

Objective 1
  Now Try
  1a. line segment, $\overline{EF}$
  1b. ray, $\overrightarrow{CD}$
  1c. line, $\overleftrightarrow{RS}$

  Practice Exercises
  1. ray, $\overrightarrow{CB}$
  3. line segment, $\overline{KL}$

Objective 2
  Now Try
  2a. intersecting
  2b. parallel

  Practice Exercises
  5. parallel

Objective 3
  Now Try
  3. ∠VSW or ∠WSV

  Practice Exercises
  7. ∠COD
  9. ∠MON

Answers

Objective 4
  Now Try
4a. straight   4b. obtuse   4c. acute

4d. right

  Practice Exercises
11. obtuse

Objective 5
  Now Try
5a. intersecting   5b. perpendicular

  Practice Exercises
13. perpendicular   15. intersecting

Objective 6
  Now Try
  6. $\angle SRT$ and $\angle TRU$; $\angle URV$ and $\angle VRW$

  8. $\angle OQP$ and $\angle PQN$; $\angle RST$ and $\angle BMC$; $\angle OQP$ and $\angle BMC$, $\angle PQN$ and $\angle RST$

  Practice Exercises
17. 164°

Objective 7
  Now Try
10. $\angle RPS \cong \angle QPT$ and $\angle QPR \cong \angle TPS$

  Practice Exercises
19. $\angle CAD \cong \angle BAE$ 21. 73° and 107°

Objective 8
  Now Try
13a. $\angle 5$ and $\angle 6$; $\angle 1$ and $\angle 2$; $\angle 7$ and $\angle 8$; $\angle 3$ and $\angle 4$

13b. $\angle 3$ and $\angle 6$; $\angle 7$ and $\angle 2$

  14. $m\angle 1 = 55°$; $m\angle 2 = 55°$; $m\angle 3 = 125°$; $m\angle 4 = 125°$
     $m\angle 5 = 125°$; $m\angle 6 = 125°$; $m\angle 7 = 55°$; $m\angle 8 = 55°$

  Practice Exercises
  23. corresponding angles: $\angle 1$ and $\angle 5$; $\angle 2$ and $\angle 6$; $\angle 3$ and $\angle 7$; $\angle 4$ and $\angle 8$
      alternate interior angles: $\angle 4$ and $\angle 6$; $\angle 3$ and $\angle 5$
      $m\angle 1 = 37°$; $m\angle 2 = 143°$; $m\angle 3 = 37°$; $m\angle 4 = 143°$
      $m\angle 5 = 37°$; $m\angle 6 = 143°$; $m\angle 7 = 37°$; $m\angle 8 = 143°$

Answers

6.6   Geometry Applications: Congruent and Similar Triangles

Key Terms
1. congruent triangles   2. congruent figures   3. similar triangles
4. similar figures

Objective 1
  Now Try
  1. ∠1 and ∠5; ∠2 and ∠4; ∠3 and ∠6; $\overline{ST}$ and $\overline{VX}$; $\overline{RS}$ and $\overline{WV}$; $\overline{RT}$ and $\overline{WX}$
  Practice Exercises
  1. ∠1 and ∠4; ∠2 and ∠5; ∠3 and ∠6; $\overline{DF}$ and $\overline{XZ}$; $\overline{DE}$ and $\overline{XY}$; $\overline{EF}$ and $\overline{YZ}$

Objective 2
  Now Try
  2a. SAS

  Practice Exercises
  3. ASA   5. SSS

Objective 3

Objective 4
  Now Try
  3. 12.75 yd

  Practice Exercises
  7. $m = 90$; $r = 21$

Objective 5
  Now Try
  5. 21 ft

  Practice Exercises
  9. 36 m   11. 10 m

Chapter 7 PERCENTS

7.1 The Basics of Percent

Key Terms
1. ratio
2. percent
3. decimals

Objective 1
  Now Try
1a. 10%
1b. 18%

  Practice Exercises
1. 68%
3. 45%

Objective 2
  Now Try
2a. 0.23
2b. 0.08
2c. 0.153
2e. 3.02
3a. 0.48
3d. 0.008

  Practice Exercises
5. 3.10

Objective 3
  Now Try
4a. 43%
4e. 290%
4b. 75.1%

  Practice Exercises
7. 20%
9. 493%

Objective 4
  Now Try
6a. $\frac{57}{200}$
5c. $1\frac{3}{4}$

  Practice Exercises
11. $\frac{11}{60}$

Objective 5
  Now Try
7b. $93\frac{3}{4}\%$

  Practice Exercises
13. 94%
15. 85.3%

Objective 6
  Now Try
8a. $6.15
9a. $3.65

  Practice Exercises
17. $520

Answers

7.2 The Percent Proportion

Key Terms
1. whole
2. part
3. percent proportion

Objective 1
  Now Try
1a. 59
1b. 7.75
1c. unknown
2a. 650
2b. unknown
2c. 3280
3a. unknown
3b. 950
3c. 2650

  Practice Exercises
1. 250%; unknown; $50
3. $7\frac{3}{4}$%; $895; unknown

Objective 2
  Now Try
4. 2470
5. 4%
6. 150 cars

7b. $94

  Practice Exercises
5. 141.0%

7.3 The Percent Equation

Key Terms
1. percent
2. percent equation

Objective 1
  Now Try
1a. $50
1b. 15 hr
1c. 25 pounds

  Practice Exercises
1. 25 minutes
3. $2000

Objective 2
  Now Try
2a. $93.20
3a. $5.49
3b. 0.307 mile

  Practice Exercises
5. 492 televisions

Objective 3
  Now Try
6a. 300 tons
4a. $275.90
5a. 220%

  Practice Exercises
7. 75%
9. 400 magazines

Answers

7.4 Problem Solving with Percent

Key Terms
1. percent equation
2. percent of increase or decrease
3. percent proportion

Objective 1
Now Try
4. 165%
5. 77 micrograms

Practice Exercises
1. 782 members
3. 17.3%

Objective 2
Now Try
6. 125%
7. 28.3%

Practice Exercises
5. 28.6%

7.5 Consumer Applications: Sales Tax, Tips, Discounts, and Simple Interest

Key Terms
1. interest rate
2. interest
3. interest formula
4. simple interest
5. principal
6. sales tax
7. discount
8. tax rate

Objective 1
Now Try
1. Tax: $6.23; Total: $95.23

Practice Exercises
1. $22.75; $372.75
3. 5%

Objective 2
Now Try
3a. Est: $3; Exact: $3.56

Practice Exercises
5. $9.00; $9.58; $12.00; $12.77

Objective 3
Now Try
4. $9.98

Practice Exercises
7. $30; $170
9. $36.18

Objective 4
Now Try
7. $29.75; $869.75

Practice Exercises
11. $195; $975

Answers

Chapter 8 MEASUREMENT

8.1 Problem Solving with U.S. Measurements Units

Key Terms
1. metric system
2. unit fractions
3. U.S. measurement units

Objective 1
  Now Try
1a. 1 lb
1b. 4 qt

  Practice Exercises
1. 2000
3. 8

Objective 2
  Now Try
2a. 210 inches
2c. 1.25 or $1\frac{1}{4}$ minutes
2d. $6.3\overline{3}$ or $6\frac{1}{3}$ hours

  Practice Exercises
5. 5 gallons

Objective 3
  Now Try
3a. 432 inches
3b. $\frac{1}{3}$ ft
5b. 7200 minutes

  Practice Exercises
7. 3.5 or $3\frac{1}{2}$ gallons
9. 3.75 or $3\frac{3}{4}$ pounds

Objective 4
  Now Try
6a. $6.36 per pound

  Practice Exercises
11. $26\frac{1}{4}$ qt

8.2 The Metric System—Length

Key Terms
1. prefix
2. metric conversion line
3. meter

Objective 1
  Now Try
1a. mm
1b. km
1c. cm

Answers

Practice Exercises
1. m
3. km

Objective 2
 Now Try
2a. 5.4 km
2b. 76 mm

 Practice Exercises
5. 0.45 km

Objective 3
 Now Try
3a. 675 mm
3b. 0.9865 km
3c. 431 cm

4a. 23,000 mm

 Practice Exercises
7. 19.4 mm
9. 0.000035 cm

8.3 The Metric System—Capacity and Weight (Mass)

Key Terms
1. gram
2. liter

Objective 1
 Practice Exercises
1. mL
3. mL

Objective 2
 Now Try
2a. 0.973 L
2b. 3850 mL

 Practice Exercises
5. 836,000 L

Objective 3
 Now Try
3a. 48 kg
3b. 590 g
3c. 3 mg

 Practice Exercises
7. mg
9. g

Objective 4
 Now Try
4a. 3720 mg
4b. 0.084 kg

 Practice Exercises
11. 760 g

Answers

Objective 5
  Now Try
  5a. 125 mL   5b. 1.19 kg

  Practice Exercises
  13. L   15. cm

8.4 Problem Solving with Metric Measurement

Key Terms
  1. gram   2. meter   3. liter

Objective 1
  Now Try
  2. 200 g   3. 5600 mL

  Practice Exercises
  1. $12.99   3. 12 pills

8.5 Metric–U.S. Measurement Conversions and Temperature

Key Terms
  1. Celsius   2. Fahrenheit

Objective 1
  Now Try
  1. 42.3 ft   2a. 21.3 lb   2b. 75.8 L

  Practice Exercises
  1. 468.5 km   3. 737.1 g

Objective 2
  Practice Exercises
  5. 37°C

Objective 3
  Now Try
  4. 50°C   5. 302°F

  Practice Exercises
  7. 17°C   9. 204°C

Chapter 9 GRAPHS AND GRAPHING

9.1   Problem Solving with Tables and Pictographs

Key Terms
1. pictograph
2. table

Objective 1
   Now Try
   1a. 68%
   1b. United
   2a. $26.50
   2b. $31.75

   Practice Exercises
   1. 177 calories
   3. 486 calories

Objective 2
   Now Try
   3a. 45 million
   3b. 45 million

   Practice Exercises
   5. 3 million people

9.2   Reading and Constructing Circle Graphs

Key Terms
1. circle graph
2. protractor

Objective 1
   Practice Exercises
   1. $10,400
   3. $400

Objective 2
   Now Try
   3. $142,500

   Practice Exercises
   5. $95,000

Objective 3
   Now Try
   4a. mysteries: 108°; biographies: 54°; cookbooks: 36°; romance novels: 90°; science: 54°; business: 18°

Answers

4b.
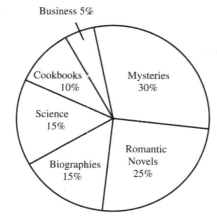

Practice Exercises

7. parts: 5%; hand tools: 20%; bench tools: 25%; brass fittings: 35%; hardware: 15%

9.
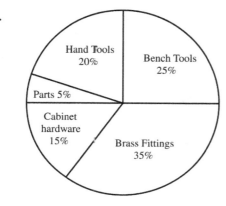

9.3 Bar Graphs and Line Graphs

Key Terms
1. line graph
2. double-bar graph
3. bar graph
4. comparison line graph

Objective 1
Now Try
1. 1800 students

Practice Exercises
1. 1600 students
3. 400 students

Objective 2
Now Try
2a. 250 female seniors
2b. 500 male freshmen

Practice Exercises
5. sophomores

Objective 3
Now Try
3a. $60   3b. December

Practice Exercises
7. September   9. $30

Objective 4
Now Try
4a. $1,500,000   4b. $4,000,000

Practice Exercises
11. $2,000,000

9.4 The Rectangular Coordinate System

Key Terms
1. ordered pair
2. horizontal axis; *x*-axis
3. vertical axis; *y*-axis
4. paired data
5. coordinate system
6. quadrants
7. origin
8. coordinates

Objective 1
Now Try
1a. and 1b.

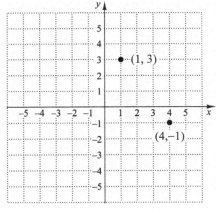

2a. and 2b.

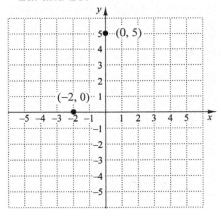

3a. (−4, −4)   3b. (1, −3)   3c. (2, 4)

3d. (0, −5)

Answers

Practice Exercises
1.
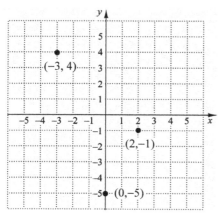

Objective 2
 Now Try
 4a. Quadrant I   4b. Quadrant II   4c. not in any quadrant

 Practice Exercises
 3. Quadrant IV   5. none

9.5 Introduction to Graphing Linear Equations

Key Terms
1. graph a linear equation   2. slope

Objective 1
 Now Try
 1. 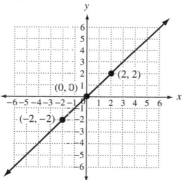   2.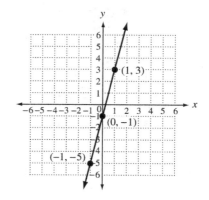

360   Copyright © 2018 Pearson Education, Inc.

4.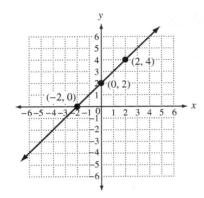

Practice Exercises

1.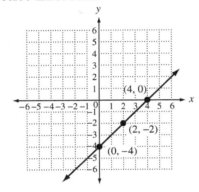

3.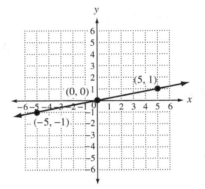

Objective 2
  Now Try
  5. positive; increases

  Practice Exercises
  5. negative

Answers

Chapter 10 EXPONENTS AND POLYNOMIALS

10.1 The Product Rule and Power Rules for Exponents

Key Terms
1. power
2. exponential expression
3. base

Objective 1
  Now Try
2a. base: 2; exponent: 6; value: 64
2b. base: 2; exponent: 6; value: –64
2c. base: –2; exponent: 6; value 64

  Practice Exercises
  1. $\left(\frac{1}{3}\right)^5$; $\frac{1}{243}$ 3. –6561; base: 3; exponent: 8

Objective 2
  Now Try
  4. $18x^{11}$

  Practice Exercises
  5. $8c^{15}$

Objective 3
  Now Try
  5d. n^{30}

  Practice Exercises
  7. 7^{12} 9. 2^{21}

Objective 4
  Now Try
  6b. $7p^3q^3$

  Practice Exercises
  11. $27a^{12}b^3$

Objective 5
  Now Try
  7a. $\frac{125}{216}$ 6c. $200x^8y^6$

  Practice Exercises
  13. $-\frac{8x^3}{125}$ 15. $-\frac{128a^7}{b^{14}}$

10.2 Integer Exponents and the Quotient Rule

Key Terms
1. power rule for exponents
2. base; exponent
3. product rule for exponents

Objective 1
  Now Try
1a. 1   1e. 1   1f. −1   1c. 88

  Practice Exercises
1. −1   3. 0

Objective 2
  Now Try
2b. $\dfrac{1}{27}$ 2c. $\dfrac{1}{8}$

  Practice Exercises
5. $\dfrac{1}{n^9}$

Objective 3
  Now Try
3a. 9   3b. $\dfrac{1}{16}$ 3d. z^{10}

  Practice Exercises
7. 13^7 9. $\dfrac{1}{z^{36}}$

Objective 4
  Now Try
4a. 10^1 or 10   4b. $\dfrac{1}{10^{14}}$ 4c. $\dfrac{1}{y^8}$

  Practice Exercises
11. p^{12}

Answers

10.3 An Application of Exponents: Scientific Notation

Key Terms
1. scientific notation
2. power rule
3. quotient rule

Objective 1
  Now Try
1a. 6.8×10^8
1b. 4.771×10^{10}
1c. 9.991×10^0
1d. 4.63×10^{-2}

  Practice Exercises
1. 2.3651×10^4
3. -2.208×10^{-4}

Objective 2
  Now Try
2a. 835,000
2c. −0.000164

  Practice Exercises
5. 0.0064

Objective 3
  Now Try
3a. 2.7×10^8, or 270,000,000
3b. 3×10^{-8}, or 0.00000003

  Practice Exercises
7. 2.53×10^2, or 253

Objective 4
  Now Try
5. 5.45×10^3 kg/m^3

  Practice Exercises
9. 2.23×10^9 (rounded) or about 2,230,000,000 pretzels
11. 3.34×10^{12} (rounded) or about 3,340,000,000,000 kg/m^3

Answers

10.4 Adding and Subtracting Polynomials

Key Terms
1. degree of a term
2. descending powers
3. trinomial
4. polynomial
5. monomial
6. degree of a polynomial
7. binomial

Objective 1
Now Try
1a. $5x^4$
1b. $-8x^7$
1c. $15m^3 + 29m^2$
1d. $5p^2q$

Practice Exercises
1. $-3z^3$
3. $8c^3 - 8c^2 - 6c + 6$

Objective 2
Now Try
2a. $8x^3 + 4x^2 + 6$; degree 3; trinomial
2b. $4x^5$; degree 5; monomial

Practice Exercises
5. $-5.7d^8 - 1.1d^5 + 3.2d^3 - d^2$; degree 8; none of these

Objective 3
Now Try
3. 1285; 1357

Practice Exercises
7. a. −51; b. 74
9. a. 29; b. 39

Objective 4
Now Try
4b. $9x^4 + 7x^2 + 7x - 6$
5a. $4x^3 + x + 12$

Practice Exercises
11. $4x^2 + 6x - 18$

Objective 5
Now Try
6b. $-11x^3 - 7x + 1$

Practice Exercises
13. $-8w^3 + 21w^2 - 15$
15. $3x^3 + 7x^2 + 2$

Answers

10.5 Multiplying Polynomials: An Introduction

Key Terms
1. monomial
2. distributive property
3. trinomial
4. polynomial
5. binomial

Objective 1
  Now Try
  1a. $32x^4 + 64x^3$
  1b. $-35m^8 + 42m^7 - 28m^6 + 7m^5$

  Practice Exercises
  1. $35z^4 + 14z$
  3. $-6y^5 - 9y^4 + 12y^3 - 33y^2$

Objective 2
  Now Try
  2a. $x^2 + 3x - 54$
  2b. $4x^7 - 2x^5 + 37x^4 - 18x^2 + 9x$
  3. $28x^4 - 33x^3 + 51x^2 + 17x - 15$

  Practice Exercises
  5. $6m^5 + 4m^4 - 5m^3 + 2m^2 - 4m$